Human Flourishing in the Age of AI

AF688955

Reclaiming Meaning, Dignity, and Purpose as Intelligence Becomes Automated

By Laura Lee

Lauxon Publishing, LLC

Published by Lauxon Publishing, LLC

For permission requests, please contact:
Email: info@Lauxonpublishing.com
Website: www.Lauxonpublishing.com

Introduction:

Why Human Flourishing Matters Now More Than Ever

Artificial intelligence is changing everything: jobs are automated, creative processes are augmented, relationships are mediated, and identity itself is being reframed in the digital mirror. The proliferation of intelligent systems is not just a technological revolution; it is a human one. How we respond will shape the next generation, the structure of our societies, and the ethical landscape of our time.

The challenge is complex. On one hand, AI can liberate humans from repetitive labor, expand knowledge, and provide tools for creativity and problem-solving. On the other hand, it can exacerbate inequality, reinforce bias, and disconnect us from ourselves and each other. Flourishing in this context requires more than adaptation—it requires intentional cultivation of the qualities that define our humanity.

This book is organized to help readers navigate the complexity of this moment. It begins with the structural and societal shifts that AI brings, moves through the personal, ethical, and relational dimensions of human life, explores collaboration with intelligent systems, and concludes with guidance for shaping a future that prioritizes collective well-being, creativity, and purpose. Demographics, class, immigration, and generational perspectives are considered throughout, recognizing that AI affects communities differently.

By the end of this journey, the goal is clear: to provide a roadmap for human flourishing in an age of AI, emphasizing ethics, creativity, emotional intelligence, and shared purpose. We are not writing a manual on resisting technology; we are writing a guide to integrating it wisely, intentionally, and humanely. Flourishing is not optional—it is the challenge and opportunity of our time.

Table of Contents:

Part I — The Great Shift

Chapter 1: The Moment Humanity Didn't See Coming
Artificial intelligence is not a future event but a present force reshaping everyday life.

Chapter 2: From Tools to Teammates
Machines are no longer passive instruments but active collaborators in human systems.

Chapter 3: Automation Anxiety and the Fear Beneath It
Economic fear often masks deeper concerns about identity, worth, and belonging.

Chapter 4: When Intelligence Is No Longer Rare
The abundance of machine intelligence challenges how humans define value.

Part II — What It Means to Stay Human

Chapter 5: Humanity Beyond Productivity
Human worth must be disentangled from economic output in an automated world.

Chapter 6: Creativity After the Algorithm
Human imagination evolves rather than disappears when machines can create.

Chapter 7: Emotional Intelligence as a Human Frontier
Empathy, care, and emotional presence become more central, not less.

Chapter 8: Relationships in a Mediated Reality
AI reshapes how humans connect, love, and form community.

Chapter 9: Spirituality in an Engineered World
AI forces ancient questions about meaning, soul, and transcendence back into focus.

Part III — Human–Machine Collaboration

Chapter 10: The Myth of Man Versus Machine
The future favors partnership over competition between humans and AI.

Chapter 11: Augmented Intelligence, Not Artificial Replacement
AI amplifies human capability when designed with intention.

Chapter 12: Decision-Making in the Age of Algorithms
Human judgment remains essential even as machines optimize choices.

Chapter 13: Trust, Transparency, and Control
Healthy collaboration requires understanding how machines think and act.

Part IV — Ethics, Power, and Inequality

Chapter 14: Who Owns the Machines That Shape Reality
Control over AI systems determines who benefits and who is left behind.

Chapter 15: Bias in Code and the Reinvention of Discrimination
AI can reinforce inequality unless consciously designed to resist it.

Chapter 16: Black, White, Brown, and Invisible
Racialized communities experience AI differently based on history and access.

Chapter 17: AI and the Global South
Emerging economies face both unprecedented opportunity and new exploitation.

Chapter 18: Class in an Automated Economy
Lower-income and middle-class workers face disproportionate disruption.

Chapter 19: Immigration, Borders, and Digital Labor
AI reshapes migration, work visas, and global competition for talent.

Part V — Work, Wealth, and the End of Necessity

Chapter 20: The Automation of Labor
Many jobs will disappear, but human contribution will not.

Chapter 21: When Work Becomes Optional
AI raises the realistic possibility of a post-work society.

Chapter 22: Universal Basic Income and Economic Dignity
New economic models attempt to separate survival from employment.

Chapter 23: Purpose Without Paychecks
Humans must redefine meaning beyond traditional labor.

Chapter 24: The Psychological Impact of Jobless Abundance
Freedom from work can liberate or destabilize the human psyche.

Part VI — Skills for the Next Decade

Chapter 25: Learning How to Learn Again
Adaptability becomes the most valuable skill of all.

Chapter 26: Critical Thinking in a World of Instant Answers
Discernment matters more when information is effortless.

Chapter 27: Ethical Reasoning as a Practical Skill
Moral judgment becomes essential in AI-mediated decisions.

Chapter 28: Cultural Intelligence and Global Awareness
AI connects societies while amplifying cultural misunderstandings.

Chapter 29: Creativity, Curiosity, and Courage
Human originality thrives where machines cannot feel or wonder.

Part VII — The Next Generation

Chapter 30: Children Growing Up with Thinking Machines
AI will shape how the next generation learns, plays, and imagines.

Chapter 31: Education Beyond Memorization
Schools must prepare students for a world where answers are instant.

Chapter 32: Identity Formation in an Algorithmic Mirror
Young people will define themselves in constant dialogue with machines.

Chapter 33: Digital Companions and Synthetic Relationships
AI friends and mentors raise new emotional and ethical questions.

Chapter 34: Teaching Wisdom, Not Just Knowledge
The future demands character development as much as technical skill.

Part VIII — Choosing a Flourishing Future

Chapter 35: Designing AI for Human Good
Technology reflects the values of those who build it.

Chapter 36: Governance, Responsibility, and Global Cooperation
Unchecked AI power requires coordinated human oversight.

Chapter 37: Slowing Down in a World That Accelerates
Human well-being depends on resisting constant optimization.

Chapter 38: Reclaiming Community and Shared Meaning
Flourishing is collective, not individual.

Chapter 39: Hope Without Naivety
The future demands realism grounded in possibility.

Chapter 40: Becoming More Human Than Ever
AI's greatest gift may be forcing humanity to rediscover itself.

Prologue:

The Dawn of a New Humanity

The world is changing faster than most of us can comprehend. Artificial intelligence is no longer a distant possibility or a futuristic concept—it is already reshaping our daily lives, our work, our relationships, and even our sense of self. Machines can analyze vast amounts of data, compose music, generate art, and make decisions that were once the exclusive domain of humans. In this context, the question is no longer whether AI will change humanity, but how humanity will rise to meet AI.

The pace of technological advancement can be disorienting. Algorithms make recommendations before we even realize we need them, automation removes repetitive labor, and AI-generated content floods our attention. These changes bring immense opportunity, but they also demand reflection, ethical awareness, and intentional action. We must consider what it means to thrive—not merely survive—in a world increasingly shared with intelligent machines.

This book is an invitation to explore those questions deeply. It is a call to reclaim what is uniquely human while collaborating with technology responsibly. It asks: How can we cultivate empathy, creativity, ethical judgment, and purpose in a society where intelligence is abundant but wisdom remains rare? How can we ensure that innovation enhances human life for all, rather than concentrating power or eroding dignity?

The chapters ahead examine the intersection of AI and humanity from multiple angles: our work, our relationships, our ethics, and our communities. They explore the opportunities for human flourishing that emerge when we choose to reflect, collaborate, and act intentionally in an AI-enhanced world. This is not a manual for avoiding technology—it is a guide to becoming more human than ever.

Above all, this book asserts a simple but profound truth: AI challenges humanity not by replacing it, but by revealing what is most precious, powerful, and enduring about being human. In the following pages, we will navigate fear, hope, possibility, and responsibility together, imagining a future where humans thrive alongside intelligent machines.

Part I — The Great Shift

Chapter . 1

The Moment Humanity Didn't See Coming

Artificial intelligence is not a future event but a present force reshaping everyday life.

Artificial intelligence did not arrive with a single dramatic announcement or a visible turning point. It seeped quietly into daily life through recommendations, predictions, and invisible optimizations. People woke up one morning to discover that machines were deciding what they watched, what they bought, and sometimes what they believed. Unlike past technological revolutions, this one did not demand attention through noise or spectacle. It asked instead for permission through convenience. Humanity accepted without fully understanding the cost.

For decades, artificial intelligence was framed as a distant possibility reserved for science fiction and research labs. Popular imagination pictured sentient robots or hostile machines announcing their arrival with menace. In reality, AI appeared as something far more ordinary and far more persuasive. It lived inside phones, search engines, cameras, and scheduling tools. It learned silently from human behavior rather than declaring independence. By the time it became noticeable, it was already indispensable.

This quiet integration blurred the boundary between human decision-making and machine guidance. People rarely noticed when a choice had been shaped rather than freely made. Navigation systems replaced memory, recommendation engines replaced curiosity, and automation replaced reflection. Each surrender felt small and harmless in isolation. Collectively, these shifts altered how humans related to knowledge, effort, and responsibility. The transformation happened not through force but through trust.

The speed of this change caught institutions unprepared and individuals unaware. Legal systems, educational models, and moral frameworks were built for a slower world. They assumed human cognition would remain the central driver of progress. Artificial intelligence disrupted that assumption without asking permission. It outpaced regulation and philosophical reflection alike. Society began reacting only after dependence was already established.

What makes this moment historically unique is not intelligence itself but its scale and accessibility. Intelligence, once rare and tied to human development, became abundant and instantly available. A tool that could reason, predict, and create was suddenly in everyone's pocket. This abundance changed the meaning of expertise and effort. It also forced humanity to confront an uncomfortable question. If intelligence is no longer scarce, what then defines human value?

The moment humanity did not see coming was not marked by collapse or catastrophe but by normalization. Artificial intelligence became ordinary before it became controversial. People adjusted their habits faster than their values. Convenience slowly replaced deliberation as the dominant cultural priority. Systems optimized for speed began shaping expectations of human performance. The quietness of this shift made it easy to underestimate its significance.

Workplaces were among the first spaces to feel this transformation. Algorithms screened resumes, scheduled labor, and evaluated productivity with increasing authority. Decisions once made by managers or peers were delegated to systems trained on historical data. For many workers, especially those in lower-income and middle-class roles, these changes arrived without explanation or consent. Automation felt impersonal and unavoidable. The human cost was often framed as inefficiency rather than displacement.

Education followed a similar path of subtle disruption. Tools designed to assist learning gradually began to replace core cognitive tasks. Students relied on machines to summarize, calculate, and even compose. Memorization lost relevance while speed gained prestige. Teachers struggled to adapt curricula designed for a pre-algorithmic age. The purpose of education itself became unclear in a world where answers were instant.

Cultural life also shifted under the influence of artificial intelligence. Music, art, and writing were no longer exclusively human expressions. Algorithms learned patterns of beauty and reproduced them at scale. Audiences often could not distinguish between human-made and machine-generated work. This blurred distinction challenged long-held beliefs about creativity and originality. The question was no longer whether machines could create, but whether creation still defined humanity.

Despite its profound impact, this transition rarely inspired collective reflection. News cycles treated AI as either a novelty or a threat, rarely as a mirror. Public discourse oscillated between hype and fear, skipping the deeper conversation about meaning. Many sensed something fundamental was changing but lacked language to describe it. The absence of shared understanding intensified anxiety and confusion. Humanity moved forward without fully agreeing on where it was going.

Beneath the surface of technological progress lay a more intimate disruption. Artificial intelligence began to reshape how people understood themselves. Identity, once formed through work, skill, and contribution, became unstable. When machines could perform cognitive tasks once considered uniquely human, confidence wavered. Many experienced a quiet erosion of self-worth. This loss was rarely named but deeply felt.

For generations, societies taught that effort led to mastery and mastery led to dignity. AI complicated that story by rewarding outcomes rather than process. A machine could arrive at an answer without struggle, memory, or growth. Human effort suddenly seemed inefficient by comparison. The cultural admiration for hard work began to weaken. In its place emerged a subtle pressure to keep up with systems that never tired.

This pressure affected different communities in unequal ways. Those with access to technology gained leverage, while those without faced increased marginalization. Racialized communities, immigrants, and workers in precarious roles often became subjects of algorithmic judgment rather than beneficiaries of its power. Historical inequalities were encoded into data and repeated at scale. The promise of neutrality masked the reality of bias. Technology inherited the moral failures of its creators.

At the same time, wealth concentrated around those who owned and controlled intelligent systems. Power shifted from labor to capital with unprecedented speed. Entire industries reorganized around automation rather than human skill. The middle ground between innovation and exploitation narrowed. Many people sensed they were being left behind by forces they did not choose. The future felt decided elsewhere.

Yet even as disruption spread, resistance remained muted. People adapted individually rather than collectively. They learned new tools, optimized themselves, and hoped to remain relevant. Few questioned whether relevance should be the measure of a human life. The deeper issue was not technological capability but philosophical direction. Humanity was moving fast without asking what it was moving toward.

Amid this rapid shift, human attention became a scarce resource. Machines captured focus through notifications, personalized feeds, and optimized engagement. Time once devoted to reflection or exploration was absorbed by systems designed to anticipate desires. Minds were trained to respond rather than to inquire. Curiosity and patience, once central to learning, were quietly eroded. The human rhythm of observation, contemplation, and creation was disrupted without overt awareness.

The workplace evolved into a site of constant measurement and optimization. Employees were no longer judged solely by output but by metrics extracted from every interaction. AI monitored performance, collaboration, and even emotional cues. This created an environment where stress and anxiety became normative. Human beings were assessed as datasets, and their value often felt contingent on algorithmic approval. Autonomy was diminished, leaving a sense of invisible control.

Social interactions also changed under the influence of AI. Digital mediation altered how relationships were initiated, maintained, and evaluated. Recommendation systems suggested friends, partners, and collaborators based on data-driven predictions. Conversations became curated experiences, filtered to maximize comfort and minimize conflict. While efficiency improved, authenticity and spontaneity suffered. Emotional lives were increasingly guided by machines that knew human preferences better than humans themselves.

In parallel, creative expression faced profound challenges. AI-generated music, visual art, and literature raised questions about originality and ownership. The tools that promised democratization also threatened devaluation. Artists struggled to distinguish their work in a world where replication was instantaneous. Audiences were confronted with an abundance of content, making attention fleeting and competition relentless. Yet, within this challenge lay potential for a new synthesis of human imagination and machine capability.

Despite these pressures, moments of human agency persisted. People found ways to assert choice and meaning within algorithmic constraints. They sought creative outlets, community connection, and personal reflection. Even small acts of resistance or mindfulness became declarations of autonomy. Humanity retained the capacity to decide how to interact with AI rather than passively accept it. The struggle was ongoing, but it demonstrated that the present moment, though unanticipated, was not devoid of possibility.

The realization that humanity did not foresee this moment brings both urgency and opportunity. It is a reminder that technological progress is not neutral—it reflects human priorities, biases, and imagination. Every choice in design, deployment, and adoption shapes society in profound ways. To ignore these decisions is to participate passively in a world we may not fully understand. Yet awareness alone is transformative; it opens the door for intentional action. The present moment, though unplanned, can become a space for conscious shaping of the future.

As artificial intelligence permeates life, humans are challenged to reconsider what it means to live purposefully. Value can no longer be defined solely by productivity or output. Meaning arises from connection, creativity, ethical engagement, and care. AI can enhance these pursuits but cannot replace them. Humans are called to cultivate capacities that remain uniquely theirs: empathy, wisdom, and moral imagination. Flourishing in this era depends on reclaiming these distinctly human domains.

Communities, too, must respond with deliberate thought. Policy, education, and public discourse must catch up to technology that evolves at breakneck speed. Inclusion and equity require proactive design, not reactive fixes. Without intentional guidance, AI risks amplifying historical inequities. Collective engagement ensures that advancements benefit all rather than concentrating advantage among the few. Society faces the choice to steward intelligence rather than be steered by it.

The personal dimension of this moment is equally pressing. Individuals must confront the subtle ways AI shapes choices, attention, and self-perception. Mindfulness, reflection, and discernment are essential skills. Humans can cultivate the ability to pause, question, and redirect influence in alignment with values. This self-awareness transforms technological dependency from passive surrender to active partnership. Flourishing begins with conscious participation.

Ultimately, the arrival of AI is not the end of humanity but a test of its resilience. The challenge is not merely to survive but to flourish in new contexts. By understanding the forces at play, nurturing human capacities, and asserting ethical agency, people can turn disruption into opportunity. The future will not be given; it must be co-created with intentionality and courage. Humanity stands at the threshold of a new chapter—one that demands both humility and vision.

Author Reflection:

Writing this chapter forced me to confront my own complicity in the unnoticed rise of AI. I realized that, like everyone else, I had grown accustomed to conveniences without questioning their consequences. Reflection became a practice not only for understanding technology but for understanding myself. The silent encroachment of AI mirrors a broader human tendency to follow ease over awareness. Recognizing this is uncomfortable, but necessary. Flourishing will require constant vigilance and conscious choice.

Inspirational Quote:

"The real question is not whether machines think but whether humans do." — B.F. Skinner

Chapter . 2

From Tools to Teammates

Machines are no longer passive instruments but active collaborators in human systems.

For centuries, humans have used tools to extend physical strength, increase efficiency, and solve problems. Hammers, plows, and levers enhanced human capability without requiring independent thought. Machines were clearly subordinate to human intention. Artificial intelligence, however, challenges that distinction. Algorithms can now analyze, predict, and act in ways that influence outcomes without direct human intervention. What was once a clear hierarchy between tool and user is becoming ambiguous.

Collaboration with machines is no longer limited to repetitive tasks. Modern AI can optimize complex systems, from supply chains to healthcare diagnostics. It can detect patterns invisible to the human eye, providing insights previously unimaginable. This collaboration extends human reach while challenging traditional authority. Decision-making becomes a shared responsibility between human judgment and machine reasoning. The nature of expertise is evolving as humans rely more on algorithmic insight.

AI teammates also alter the dynamics of creativity. Design, music, and writing are increasingly co-created with intelligent systems.

Humans contribute intuition, context, and emotion, while machines supply pattern recognition and scale. This partnership generates outcomes neither could achieve alone. It forces a reconsideration of originality and authorship. Creative processes become dialogues rather than solo endeavors.

The workplace offers vivid examples of human-machine teamwork. Employees collaborate with AI to schedule complex projects, manage logistics, and analyze data trends. Machines handle calculations, risk assessment, and forecasting. Humans provide ethical judgment, negotiation skills, and context-sensitive decision-making. The combination often outperforms either alone. Teams are no longer purely human; they are hybrid networks of thought and action.

Despite its potential, collaboration with machines introduces friction. Misalignment between human goals and algorithmic objectives can produce unintended consequences. Dependence on machine suggestions risks eroding critical thinking. Transparency in AI reasoning is often limited, creating gaps in understanding. Building trust becomes essential to effective partnership. Humans must learn to guide, question, and intervene intelligently.

Effective collaboration begins with understanding the strengths and limitations of AI. Machines excel at speed, pattern recognition, and processing vast amounts of data. Humans excel at judgment, empathy, and contextual interpretation. Combining these complementary capabilities is the foundation of a productive partnership. Yet this requires conscious effort rather than passive reliance. Without awareness, humans risk ceding too much authority to systems they do not fully understand.

The relationship between human and machine is also shaped by expectations. People often anthropomorphize AI, assuming intent or understanding where none exists. This can lead to overconfidence in automated recommendations or frustration when outcomes fall short. Educating users about the nature of algorithms is critical. Clear communication about capability, limitations, and scope builds more resilient collaborations. The challenge lies in aligning perception with reality.

In creative industries, AI teammates are redefining collaboration norms. Designers, artists, and writers work alongside generative systems to explore possibilities. Machines suggest variations, predict trends, and simulate outcomes. Humans then filter, curate, and refine the results. This iterative dialogue expands creative potential while preserving human intentionality. It also raises new questions about ownership, responsibility, and the meaning of authorship.

Healthcare is another field transformed by collaborative intelligence. Algorithms can identify anomalies in medical imaging, predict patient risk, and optimize treatment plans. Doctors, however, retain responsibility for ethical decisions, bedside manner, and nuanced judgment. The partnership enhances outcomes while demanding vigilance. Patients benefit from both the efficiency of machines and the humanity of practitioners. In this space, the synergy between human and machine is tangible and life-altering.

Education, too, experiences a profound shift with AI as a teammate. Adaptive learning platforms analyze student performance in real time. Teachers use these insights to personalize instruction and identify gaps. AI handles repetitive evaluation, freeing educators to focus on mentorship, inspiration, and critical thinking. Collaboration here is less about replacement and more about augmentation. Students gain support that is timely, data-driven, and responsive. The human touch remains irreplaceable in guiding growth.

Trust forms the cornerstone of human-machine collaboration. Humans must rely on AI to provide accurate, relevant, and timely guidance. Machines, in turn, function within the parameters set by human designers and operators. When trust breaks down, efficiency and safety suffer. Overreliance or blind faith in AI can be as dangerous as complete rejection. Cultivating discernment in collaboration is therefore both ethical and practical.

Communication between humans and AI requires new literacies. Understanding how algorithms make decisions, what data informs them, and where biases exist is essential. Users must interpret outputs critically and contextually. Machines provide information, but humans decide its meaning and application. This layered interaction creates a feedback loop of learning and adjustment. Over time, both parties—human and machine—become more attuned to each other's capabilities and limits.

Cultural differences also influence collaboration with AI. Attitudes toward authority, innovation, and risk affect how teams integrate intelligent systems. In some societies, deference to machines may be more pronounced, while in others skepticism dominates. Organizations must recognize these dynamics to foster inclusive, effective partnerships. Collaboration is as much social and cultural as it is technical. A nuanced approach ensures that AI serves collective goals rather than reinforcing existing inequities.

AI's role as a teammate challenges long-held assumptions about creativity and labor. Humans may fear that machines will diminish their unique contributions. Yet collaboration often elevates human work rather than displacing it. Routine tasks can be automated, freeing humans to focus on interpretation, strategy, and relational work. The partnership creates space for more meaningful engagement. Flourishing emerges when machines handle what they do best, and humans reclaim what only they can do.

The shift from tool to teammate also demands ethical vigilance. Machines act without consciousness, but their actions carry moral weight. Decisions embedded in algorithms reflect human choices about priorities, fairness, and value. Humans must remain accountable for the consequences of machine-mediated actions. Ethical collaboration is not optional—it is the foundation of responsible integration. Humanity's task is to guide intelligence, not simply to deploy it.

The evolving partnership between humans and machines is transforming organizational structures. Hierarchies are flattened as AI distributes information and insights across teams. Decision-making becomes more networked, relying on both human intuition and machine analysis. Leaders must balance authority with trust in technological processes. This new dynamic requires flexibility and constant recalibration. Success depends on embracing change without losing sight of human values.

Learning to collaborate effectively also reshapes professional development. Employees must acquire technical literacy alongside interpersonal and ethical skills. Understanding data, algorithms, and system design becomes as crucial as communication and empathy. Organizations that invest in this dual capacity cultivate resilience and innovation. Collaboration is not a static state but a continuous process of adaptation. Individuals who can navigate both human and machine contributions thrive in this new landscape.

The emotional dimension of teamwork is affected as well. Humans experience frustration, uncertainty, and dependency in interactions with AI. Machines, while devoid of feeling, shape human emotions through feedback and outcomes. Awareness of this influence allows humans to regulate responses and maintain agency. Emotional intelligence becomes a critical competency in guiding collaboration. By attending to both technical and emotional dynamics, teams can achieve harmony and efficiency.

AI teammates also challenge assumptions about accountability. When decisions result from joint human-machine processes, responsibility can feel diffuse. Legal, ethical, and organizational frameworks must adapt to this shared authorship of outcomes. Clear protocols, documentation, and transparency help maintain trust and integrity. Human judgment remains the ultimate arbiter, even when machines provide substantial input. Accountability is strengthened when humans embrace their guiding role rather than deflecting to technology.

Finally, collaboration with AI prompts reflection on purpose and meaning in work. Machines excel at execution but lack vision, empathy, and ethical reasoning. Humans define the “why” behind every decision, infusing labor with intention and value.

Recognizing this distinction reframes anxiety about displacement as an opportunity for enrichment. The partnership invites humans to focus on creativity, connection, and stewardship. Flourishing depends on embracing this shift consciously and responsibly.

The full realization of AI as a teammate transforms not only processes but mindsets. Humans begin to see machines not as competitors but as collaborators that extend capability. This perspective fosters innovation, experimentation, and a willingness to explore new solutions. Resistance gives way to curiosity, and fear is replaced by intentional engagement. By embracing partnership, humans reclaim agency over technology rather than being passively shaped by it. The key lies in maintaining awareness of what only humans can contribute.

Collaboration also deepens when humans recognize the moral and ethical implications of AI decisions. Machines follow the rules embedded in their design but cannot weigh competing values or consider human consequences independently. Human guidance ensures that technological power aligns with social responsibility. The partnership thus becomes a moral endeavor as much as a technical one. Flourishing in this era requires integrating intelligence with conscience. The most effective teams are those that honor both dimensions.

Innovation thrives when humans and machines operate in dialogue rather than hierarchy. Creative problem-solving, strategic planning, and ethical decision-making all benefit from this interplay. Human intuition is informed and expanded by machine insight. Feedback loops accelerate learning and reduce error. Collaboration encourages shared ownership of outcomes, blending the strengths of both participants. The result is a synthesis greater than the sum of its parts.

Yet, the partnership is not without tension. Misalignment between human intention and machine output can generate frustration and error. Trust must be continuously nurtured, and communication refined. Humans need to remain vigilant, critically evaluating machine suggestions. Blind reliance undermines the very benefits collaboration seeks to provide. Awareness, discernment, and reflection are the tools for maintaining a healthy, productive relationship. In this space, humans remain central, not obsolete.

Ultimately, AI as a teammate challenges humanity to rethink the meaning of labor, creativity, and responsibility. Machines can amplify what humans do best but cannot replace conscience, empathy, or purpose. Embracing this partnership requires courage, humility, and ethical commitment. It invites reflection on what it means to collaborate across species of intelligence—biological and artificial. Those who learn to navigate this new landscape with wisdom will thrive. The promise of human-machine teamwork lies not in substitution, but in shared flourishing.

Author Reflection:

Writing this chapter made me appreciate the subtlety of collaboration. Too often, I had framed AI as a tool or threat, overlooking the potential for dialogue and partnership. Reflecting on real-world examples clarified that humans are not being replaced—they are being extended. Recognizing the unique contributions humans bring to the table is both empowering and humbling. Flourishing depends not on resisting AI, but on engaging it consciously and ethically. Partnership, I realized, is an act of responsibility and imagination.

Inspirational Quote:

"*Technology is a word that describes something that doesn't work yet.*" — Douglas Adams

Chapter . 3

Automation Anxiety and the Fear Beneath It

Economic fear often masks deeper concerns about identity, worth, and belonging.

The rise of automation has stirred anxiety that goes beyond the loss of jobs. People fear not only unemployment but also the erosion of purpose, skill, and dignity. Work has historically provided structure, social connection, and identity, and its disruption reverberates throughout life. Anxiety often manifests as worry about financial security, but its roots are deeper. At the core lies a concern about what it means to be human in a world where machines perform tasks once considered uniquely human. This unease is universal, affecting individuals across professions, cultures, and economic levels.

Economic anxiety is compounded by speed and unpredictability. Technology evolves faster than policy, regulation, or personal adaptation. Jobs that existed yesterday may vanish tomorrow, and skills that once ensured stability may no longer suffice. This temporal dissonance amplifies stress and uncertainty. Workers face a paradox: adapt continuously or risk obsolescence. The pressure to remain relevant is relentless, often leaving little space for reflection or recalibration.

Automation anxiety is not solely about survival; it is also about meaning. Humans derive satisfaction from accomplishment, mastery, and contribution. When machines take over tasks, these sources of fulfillment are threatened. Individuals may feel displaced even if their livelihood remains intact. Emotional and psychological impacts often exceed tangible economic losses. Anxiety becomes a reflection of a deeper fear: that one's value is conditional on work that machines can now perform.

The experience of automation anxiety varies across demographic lines. Lower-income workers, marginalized communities, and immigrants often face disproportionate exposure to job displacement. Access to retraining, networks, and technological resources is uneven. Racialized and socioeconomically disadvantaged groups may perceive AI as reinforcing historical inequities. Even for those in secure positions, anxiety persists in subtler forms—fear of irrelevance, obsolescence, or social comparison. The psychological dimensions of automation anxiety are intertwined with broader social and structural factors.

Acknowledging this anxiety is the first step toward mitigation. Denial or dismissal only heightens tension and vulnerability. Understanding the sources of fear—financial, social, and existential—allows for deliberate responses. Individuals can cultivate resilience through skill development, ethical awareness, and community engagement. Organizations can support workers with transparency, training, and participatory decision-making. Society benefits when fear is recognized as a guide, signaling where intervention and adaptation are necessary.

Automation anxiety also reveals the fragility of identity tied exclusively to work. For generations, people measured worth by productivity, output, and skill mastery. When machines perform these tasks, the equation shifts. Humans are forced to question what, beyond work, defines value. Parents, teachers, and mentors may struggle to convey a sense of purpose that transcends economic function. This reevaluation can be unsettling, yet it is also an opportunity to explore deeper sources of fulfillment.

The fear beneath automation anxiety is often emotional rather than rational. Humans respond to loss, unpredictability, and uncertainty more viscerally than to abstract statistics. Even individuals in professions perceived as "safe" report tension, stress, and self-doubt. Mental health consequences emerge when these anxieties are left unaddressed. Depression, burnout, and chronic stress can follow prolonged exposure to a rapidly shifting environment. Recognizing emotional reactions as legitimate provides a foundation for coping and adaptation.

Education and retraining are critical buffers against automation anxiety, but they are not sufficient alone. Learning new technical skills addresses only part of the human experience. Equally important is cultivating adaptability, resilience, and critical thinking. Emotional literacy allows individuals to navigate uncertainty with composure. Social support networks reinforce confidence and connection. A holistic approach integrates skill-building with psychological and relational development.

Organizational transparency can mitigate fear by clarifying expectations and intentions. When companies communicate how AI will augment rather than replace human roles, employees feel empowered rather than threatened. Participatory approaches allow workers to influence implementation, fostering trust and shared responsibility. Conversely, secrecy or top-down imposition amplifies anxiety. Collaboration, openness, and acknowledgment of human concerns are as vital as technical deployment. The human dimension cannot be separated from technological advancement.

Finally, automation anxiety invites reflection on societal values. Are humans valued solely for labor, or for creativity, empathy, and ethical engagement as well? Addressing this question requires cultural, institutional, and individual effort. Societies that prioritize holistic human flourishing will navigate technological disruption with resilience. Those that focus narrowly on efficiency risk widening social divides and amplifying fear. Recognizing the deeper sources of anxiety is the first step toward transforming it into a catalyst for growth.

Automation anxiety also interacts with generational perspectives. Older workers may feel their expertise and experience are devalued by machines that can learn rapidly and adapt without fatigue. Younger generations, while often more comfortable with technology, face pressure to continuously acquire new skills and demonstrate relevance. Both groups experience uncertainty, though in different ways. This generational tension can strain workplaces and social cohesion. Addressing these dynamics requires empathy, mentorship, and structured support for all age groups.

Communities are equally affected by the rise of automation. Local economies built around certain industries may face disruption, altering patterns of employment, migration, and social stability. Urban and rural regions experience different challenges; access to retraining, infrastructure, and digital literacy varies widely. Marginalized populations often bear the brunt of these changes, compounding historical inequities. Community-based initiatives that foster education, collaboration, and opportunity can buffer against widespread dislocation. Social cohesion becomes a critical factor in mitigating anxiety and fostering resilience.

Cultural narratives also shape how automation anxiety is experienced. Societies that tie identity tightly to work, achievement, or competition may see heightened fear and stress. Conversely, cultures that value community, creativity, and reflection may provide alternative frameworks for meaning. Media, literature, and public discourse reinforce these narratives, influencing collective perception. Recognizing the role of cultural framing allows for more nuanced interventions. Addressing fear is therefore both a personal and societal project.

The fear beneath automation anxiety often masks deeper existential concerns. Humans grapple with questions of purpose, relevance, and belonging in a world where machines can perform cognitive labor. The threat is not merely economic but ontological: what does it mean to be human when machines can think, decide, and create? Engaging with these questions requires philosophical, ethical, and spiritual reflection. Individuals and communities must confront these realities to navigate the emerging landscape.

Anxiety can become a guide, signaling where growth and adaptation are required.

Finally, resilience emerges as a central theme in facing automation anxiety. Individuals who cultivate adaptability, emotional intelligence, and curiosity are better equipped to thrive alongside intelligent machines. Organizations that prioritize transparency, equity, and shared purpose foster confidence and engagement. Societies that emphasize holistic human value can transform fear into opportunity. Anxiety, when acknowledged and addressed, becomes a signal for intentional action rather than a paralyzing force. Flourishing is possible even amidst uncertainty, but it requires awareness, courage, and ethical commitment.

Automation anxiety often intersects with personal and familial concerns. Parents worry not only about their own job security but also about the prospects of their children in a rapidly shifting economy. Households face stress over adapting skills, relocating for opportunities, or navigating changing social safety nets. These pressures compound, creating a ripple effect of uncertainty and emotional strain. Anxiety, in this sense, is both individual and collective. Families become microcosms of broader societal disruption.

The workplace environment can either exacerbate or alleviate anxiety. Cultures that emphasize performance metrics, constant monitoring, and competitive comparison heighten fear. Conversely, organizations that foster learning, collaboration, and open dialogue reduce tension. Human-centered policies—such as mentorship, flexible reskilling, and supportive leadership—buffer against uncertainty. Anxiety diminishes when humans perceive control and agency over their work. Machine integration, when done thoughtfully, can augment rather than threaten human well-being.

Media narratives amplify automation anxiety by framing technological progress as either utopian or apocalyptic. Sensationalized stories about mass unemployment, sentient machines, or corporate takeover dominate headlines. While dramatic accounts capture attention, they rarely provide actionable insights. Balanced, nuanced discussion is critical to help individuals contextualize change. Understanding trends without succumbing to fear empowers informed decision-making. Media literacy becomes a valuable tool in navigating the emotional landscape of AI.

The psychological dimension of automation anxiety is intertwined with self-concept. Work often defines status, competence, and contribution. When machines assume tasks previously central to identity, individuals may experience a sense of inadequacy or obsolescence. Addressing this requires cultivating sources of self-worth beyond productivity. Creativity, relationships, ethics, and personal growth become essential anchors. The human response to automation must therefore be holistic, integrating mind, emotion, and purpose.

Finally, coping strategies must balance acceptance and agency. Individuals can cultivate emotional resilience, adapt skills, and seek meaningful engagement beyond employment. Communities can provide education, social support, and inclusive opportunity. Policymakers can design frameworks that protect dignity while encouraging innovation. Anxiety, though uncomfortable, signals areas that demand attention and reflection. Recognizing fear as informative rather than paralyzing transforms it into a catalyst for proactive action. The path forward depends on conscious engagement, not avoidance.

As automation advances, fear and uncertainty will remain constant companions. Yet these emotions are not inherently negative; they can illuminate areas requiring adaptation and ethical consideration. The challenge is to confront anxiety deliberately rather than suppress it. By acknowledging both personal and societal implications, humans can regain a sense of agency. Reflection, skill-building, and connection become essential tools for navigating the changing landscape. Automation anxiety, when approached constructively, can inspire growth rather than paralysis.

Individuals must learn to distinguish between immediate threats and long-term opportunities. Machines may perform tasks once considered uniquely human, but they cannot replicate moral judgment, creativity, or emotional depth. Humans retain responsibility for guiding AI, setting ethical boundaries, and creating meaning. Anxiety can motivate preparation, experimentation, and collaboration. By channeling fear into action, humans shape a future aligned with their values. Purpose emerges when worry becomes a call to engagement rather than despair.

Societal frameworks also play a pivotal role in managing automation anxiety. Governments, educational institutions, and communities must provide structures that foster resilience and inclusion. Access to retraining, equitable opportunities, and social safety nets reduces existential stress. Transparent policies regarding AI integration reassure populations that change is guided rather than random. Collective engagement transforms anxiety into shared responsibility. Humans flourish when fear catalyzes cooperation and intentional planning.

Cultural narratives must evolve alongside technological progress. Stories of human worth tied solely to work exacerbate anxiety. Alternative narratives emphasizing adaptability, creativity, and ethical engagement cultivate confidence. Media, education, and public discourse shape perception and expectation. By reframing success beyond productivity, societies reduce the psychological burden of automation. Flourishing depends on embracing values that transcend labor metrics alone.

Ultimately, confronting the fear beneath automation anxiety is a deeply human endeavor. It demands reflection, courage, and deliberate action. Individuals must cultivate skills, relationships, and values that remain uniquely human. Communities and institutions must create environments that support adaptation, equity, and ethical guidance. By facing anxiety directly, humans transform a source of stress into a catalyst for growth and empowerment. The moment humanity did not see coming can become an opportunity for intentional, meaningful evolution.

Author Reflection:

Writing this chapter revealed how much fear is tied to identity, not just income. Automation threatens familiar markers of value, leaving emotional and existential voids. Recognizing this allows for intentional strategies to reclaim purpose. I realized that anxiety can be a guide, signaling areas where adaptation, learning, and ethical reflection are required. Flourishing is possible when fear informs conscious choices rather than dictating them. Understanding and confronting these fears is the first step toward empowerment.

Inspirational Quote:

"Do not fear the future, for it is the canvas upon which you can paint your resilience." — Anonymous

Chapter . 4

When Intelligence Is No Longer Rare

The abundance of machine intelligence challenges how humans define value.

For centuries, intelligence was a scarce and prized resource. Human societies organized themselves around knowledge, expertise, and the ability to solve complex problems. Those who possessed it held influence, status, and opportunity. Today, artificial intelligence challenges this dynamic by making certain forms of intelligence abundant and widely accessible. Machines can analyze vast datasets, recognize patterns, and make predictions with speed and precision far beyond human capacity. This shift forces humans to reconsider what it means to be intelligent—and what value remains uniquely human.

The ubiquity of machine intelligence transforms workplaces and industries. Decisions once dependent on individual expertise are now informed or dominated by algorithmic insights. Professionals in law, medicine, finance, and engineering increasingly rely on AI to guide actions. This creates a tension between human judgment and machine recommendations. The question is no longer whether intelligence exists, but how humans choose to wield it responsibly. Ethical discernment, contextual understanding, and moral reasoning remain irreplaceable human domains.

Abundance also democratizes access to intelligence in unprecedented ways. Students, entrepreneurs, and small organizations can leverage machine learning to compete with large institutions. Knowledge is no longer the exclusive domain of the elite or highly educated. Yet this accessibility comes with risks. Equal access does not ensure equal understanding or application. Humans must learn to interpret, critique, and contextualize AI output. The challenge is to combine machine intelligence with human insight to create meaningful outcomes.

Socially, the proliferation of machine intelligence reshapes how people perceive expertise. Skills once considered rare or prestigious may lose distinction. The markers of status and competence are shifting toward adaptability, creativity, and emotional intelligence. Those who excel in uniquely human capacities find new relevance. Society begins to value qualities machines cannot replicate—ethics, empathy, and imagination. Intelligence alone is no longer sufficient to define success or influence.

The abundance of machine intelligence also intensifies competition in unexpected ways. Individuals and organizations must differentiate themselves not by sheer knowledge, but by how they apply it creatively and ethically. Machines can process information faster, but they lack context, judgment, and empathy. Humans who leverage these uniquely human qualities gain new forms of influence. Collaboration between human and machine becomes a defining skill. The value of intelligence is increasingly measured by the wisdom with which it is used rather than the speed at which it is accessed.

Education faces a profound challenge in this era. Traditional models prioritized the acquisition of knowledge and the mastery of facts. When AI can supply those instantly, the role of education must evolve. Emphasis shifts to critical thinking, ethical reasoning, creativity, and emotional intelligence. Students must learn to interpret machine output and to integrate it with human values. The ability to navigate a world of abundant intelligence becomes more important than the ability to memorize it. Preparing the next generation requires cultivating judgment, curiosity, and resilience.

Inequality remains a pressing concern despite broader access to intelligence. Those with digital literacy, connectivity, and resources can harness AI more effectively. Others may find themselves left behind, reinforcing existing social and economic disparities. Machine intelligence is not automatically democratizing; it reflects the conditions of access and use. Addressing these disparities requires deliberate policy, education, and investment. Only then can the benefits of abundant intelligence be shared equitably across society.

The abundance of intelligence also reshapes creativity and problem-solving. Tasks that once required years of experience can now be accomplished with guidance from AI. This allows humans to explore higher-level thinking, experimentation, and innovation. The interplay between machine-generated insight and human interpretation fosters outcomes that neither could achieve alone. Creativity becomes a collaborative dialogue, rather than an individual endeavor. The future belongs to those who can integrate diverse sources of intelligence with imagination and purpose.

Abundant intelligence reshapes not only the workplace but social interaction and communication. Social media, recommendation systems, and predictive algorithms influence how people perceive information and make decisions. Machine intelligence can identify patterns and predict behavior at a scale impossible for humans. This introduces both opportunities for connection and risks of manipulation. Humans must develop media literacy, critical thinking, and ethical judgment to navigate these systems responsibly. The abundance of intelligence magnifies the consequences of both action and inaction.

In personal life, abundant AI challenges self-perception and identity. When machines can perform tasks once associated with intelligence—analysis, writing, and problem-solving—humans may question their own relevance. This can provoke anxiety, imposter syndrome, or diminished confidence. Yet it also presents an opportunity to focus on what machines cannot replicate: empathy, ethical reasoning, creativity, and relational intelligence. Abundance of intelligence is a call to redefine human value in ways beyond computation. How individuals respond determines whether this era becomes empowering or disempowering.

The workplace must adapt to integrate abundant intelligence effectively. Roles increasingly emphasize oversight, interpretation, and collaboration with AI. Routine cognitive labor is delegated to machines, while humans engage in judgment-driven and creative activities. Success requires balancing efficiency with meaning, technical skill with ethical discernment. Organizations that ignore this shift risk obsolescence or workforce disengagement. Abundance demands thoughtful integration rather than mere automation.

Abundant intelligence also highlights the limitations of current human systems. Institutions, laws, and social norms were designed for a world where intelligence was scarce and concentrated. Policies regarding work, privacy, accountability, and equity are now challenged by machines that operate at scale and with rapid adaptability. Societies must rethink foundational assumptions to align with this new reality. Failure to do so risks exacerbating inequality, inefficiency, and social instability. Abundant intelligence is both a tool and a test of human foresight.

The abundance of intelligence also reshapes education at every level. Schools and universities can no longer focus solely on teaching information that AI can supply instantly. Instead, education must prioritize interpretation, reasoning, ethics, and creativity. Students need to learn how to work alongside intelligent systems, discerning patterns while questioning assumptions. Teachers become facilitators of judgment, guides for ethical reflection, and cultivators of curiosity. The goal shifts from memorization to wisdom, from accumulation to synthesis. Abundance challenges educators to prepare humans for roles that cannot be automated.

Workplaces face similar imperatives. Employees must learn not only technical skills but also collaborative and ethical competencies. Organizations that integrate AI effectively cultivate environments where human judgment complements machine computation. Roles evolve to emphasize oversight, innovation, and relational intelligence. Employees must navigate ambiguity, interpret machine output critically, and engage in continuous learning. Abundance requires adaptability as much as capability, blending technical proficiency with uniquely human strengths.

Abundant intelligence also brings societal implications. Legal, economic, and ethical frameworks were designed for a world where intelligence was limited and slow. Machines now operate at scales and speeds that outpace regulation and social norms. Governments and institutions must respond proactively to maintain fairness, accountability, and public trust. Policies must address not only access but responsible use, bias mitigation, and equitable benefit distribution. Society faces both opportunity and risk in equal measure, with the balance dependent on intentional action.

The psychological impact of abundant intelligence cannot be ignored. Humans may feel diminished in comparison to machines, yet this perception overlooks the distinct qualities of human reasoning. Emotional intelligence, ethical judgment, and cultural awareness cannot be replicated by algorithms. Recognizing these strengths is essential to maintain confidence and purpose. Abundance is an invitation to redefine human value beyond computation. Flourishing emerges when humans embrace their irreplaceable capacities rather than compete against machines.

As intelligence becomes abundant, the human role in decision-making grows more nuanced. Machines can analyze, predict, and optimize, but they cannot weigh moral considerations or anticipate unintended consequences. Humans are required to set priorities, define values, and contextualize outcomes. Abundance calls for stewardship rather than replacement, placing ethical reasoning at the center of all decisions. The challenge is not scarcity of intelligence, but scarcity of wisdom and conscience. Flourishing depends on cultivating judgment alongside access to information.

The abundance of intelligence also reshapes creativity and innovation. Humans are no longer limited by the pace or capacity of their own cognition. AI can generate alternatives, propose solutions, and model complex scenarios in minutes. This allows humans to explore possibilities previously unimaginable. The partnership between human insight and machine computation expands the frontiers of problem-solving. Creativity becomes an iterative dialogue between imagination and intelligent analysis, enriching both outcomes and human skill.

Societies must grapple with inequality in the context of abundant intelligence. Access, literacy, and opportunity remain unevenly distributed, potentially reinforcing historical disparities. Equitable policies, inclusive education, and proactive infrastructure are essential to prevent concentration of advantage. Abundance alone does not guarantee fairness or prosperity; intentional human action is required. Ethical frameworks must guide not only the development but also the distribution and application of intelligence. Only then can the benefits of AI be shared widely.

Abundance also challenges identity and self-perception. When machines can perform tasks once considered benchmarks of intelligence, humans must redefine personal value. This requires a shift from measuring worth through output to measuring it through purpose, ethics, creativity, and contribution. The ability to integrate machine intelligence into human endeavors becomes a defining competency. Flourishing is not about competing with machines, but about complementing and guiding them. Humans retain centrality not through scarcity, but through vision and conscience.

Ultimately, the abundance of intelligence is both a challenge and an opportunity. It invites reflection on purpose, creativity, ethics, and human potential. Those who adapt, guide, and integrate AI thoughtfully will thrive in a world of plentiful intelligence. Machines amplify capability, but meaning remains human. Recognizing this distinction transforms anxiety into empowerment and abundance into a pathway for human flourishing. The era of abundant intelligence is not the end of human relevance—it is the beginning of a new form of human responsibility and opportunity.

Part II — What It Means to Stay Human

Chapter . 5

Humanity Beyond Productivity

Human worth must be disentangled from economic output in an automated world.

For much of modern history, human value has been tightly linked to productivity. Work defined social status, personal identity, and societal contribution. Hours spent laboring were equated with worth, and efficiency became a moral as well as economic imperative. As artificial intelligence assumes tasks once performed by humans, this traditional measure of value becomes increasingly obsolete. Machines can perform with speed, precision, and consistency far beyond human capability. The challenge is to define humanity beyond the confines of output.

Productivity alone cannot capture the full spectrum of human capacities. Creativity, empathy, ethical judgment, and relational intelligence are qualities that machines cannot replicate. These capacities become central markers of value in a world where labor is partially automated. Humans are called to cultivate skills and behaviors that go beyond efficiency. Flourishing requires embracing activities that contribute meaning, connection, and societal good. Economic metrics alone are insufficient to measure what it means to live a fulfilling life.

The shift away from productivity-centered value also necessitates psychological adaptation. Individuals accustomed to defining self-worth by accomplishments may feel disoriented. Anxiety, imposter syndrome, and existential questioning often arise when traditional benchmarks disappear. Recognizing the distinction between being and doing becomes essential. Humans must learn to measure their significance by contribution to relationships, communities, and moral outcomes. Detaching self-esteem from labor output opens pathways to deeper fulfillment.

Societal structures must evolve to support humanity beyond productivity. Education, healthcare, and cultural institutions play a role in fostering purpose, ethical awareness, and creative engagement. Policies that protect dignity, provide social support, and encourage exploration of human potential become essential. Work can still provide meaning, but it is no longer the sole arbiter of value. Communities flourish when individuals are encouraged to cultivate character, empathy, and insight alongside technical competence. Redefining worth requires collective as well as individual effort.

This shift is also deeply ethical. If humans are valued only for productivity, inequality and exclusion become morally normalized. Automation threatens to amplify this pattern, leaving those whose labor is replaced feeling dispossessed or devalued. Moving beyond productivity reframes questions of justice, opportunity, and human rights. Ethical societies must recognize the intrinsic worth of every individual, independent of economic output. Flourishing emerges when humans are affirmed for capacities that transcend computation.

The decline of productivity as the sole measure of value invites new ways of organizing work and life. Flexible schedules, remote collaboration, and reduced emphasis on billable hours can allow humans to focus on meaningful contributions. Tasks that machines can handle free humans to pursue creative, relational, and ethical endeavors. This reallocation challenges traditional workplace hierarchies and expectations. Humans must learn to navigate autonomy while maintaining purpose. Flourishing depends on cultivating responsibility and intentionality in what remains uniquely human.

Education plays a crucial role in preparing individuals for a world beyond productivity. Curriculum must shift from rote memorization and standardized metrics to critical thinking, collaboration, and moral reasoning. Students must develop the ability to contextualize information, evaluate ethical consequences, and innovate creatively. Knowledge acquisition alone is insufficient when machines can provide answers instantly. Educators must teach discernment, empathy, and problem-solving in ambiguous situations. Preparing humans to thrive requires focusing on capacities that cannot be fully automated.

The cultural dimension of value also shifts in this context. Media, narratives, and social norms have long equated work with virtue and idleness with failure. These associations need reevaluation in a world where automation performs essential functions. Humans can be virtuous, impactful, and fulfilled without constant labor. Society must celebrate contributions beyond economic output, including care, creativity, and ethical leadership. Redefining cultural metrics of worth supports mental health, social cohesion, and human dignity. Flourishing becomes a broader, more inclusive concept.

Individual psychology is deeply affected by this transition. Humans must confront internalized beliefs that equate self-worth with achievement. Detachment from productivity requires conscious effort, reflection, and support. Practices such as mindfulness, journaling, and community engagement foster self-awareness and resilience. Individuals can begin to measure value in terms of contribution to relationships, knowledge, and moral guidance rather than output alone. Emotional intelligence becomes a central tool for navigating a post-productivity identity.

Moving beyond productivity also requires rethinking the social contract. Societies have long tied social benefits, status, and recognition to labor. As automation displaces routine tasks, access to opportunity and dignity must be decoupled from work alone. Universal support structures, such as education, healthcare, and social safety nets, ensure that individuals maintain well-being independent of economic output. This allows humans to focus on meaningful engagement rather than mere survival. Flourishing emerges when basic needs are met and purpose becomes a personal and collective pursuit. The ethical imperative is to design systems that affirm intrinsic human worth.

The evolution of work itself becomes a platform for human expression. With machines handling repetitive or data-intensive labor, humans can dedicate effort to creativity, mentorship, and problem-solving. These domains offer satisfaction, social impact, and recognition that machines cannot replicate. Work becomes a choice rather than a necessity, emphasizing intention and contribution. Employees are encouraged to innovate, experiment, and engage relationally. Human agency is strengthened when labor is liberated from narrow economic metrics. Flourishing requires cultivating environments where choice aligns with purpose.

Relationships also gain new prominence in defining human value. Emotional intelligence, empathy, and collaboration are essential skills that shape family, community, and professional interactions. Machines may assist communication or provide insight, but humans remain the bearers of care, trust, and ethical judgment. Social bonds become central to identity and well-being in a world where work no longer dominates. Flourishing is deeply relational, relying on connection and shared meaning. Human worth expands beyond tasks completed to lives enriched through interaction and guidance.

Ethical reflection is essential in navigating this post-productivity era. Humans must continually assess how technology affects autonomy, fairness, and dignity. AI and automation, if deployed without ethical guidance, risk reinforcing inequality and eroding self-determination. Conscious stewardship requires integrating moral reasoning into decision-making and design. Flourishing arises from intentional choices that prioritize human value over efficiency alone. Humanity's responsibility is to ensure that technology serves as an instrument of empowerment rather than control.

Time itself takes on new significance in a world beyond productivity. Freed from constant labor demands, humans gain the opportunity to engage in reflection, learning, and exploration. Leisure is no longer indulgence but a vital space for growth, creativity, and ethical contemplation. How time is used becomes a marker of flourishing, emphasizing depth over quantity of output. Individuals can pursue projects, relationships, and personal development that enrich both themselves and their communities. The quality of human experience replaces metrics of efficiency as a measure of value.

Mental health becomes an essential dimension of flourishing. The detachment of worth from productivity can initially provoke uncertainty and anxiety. Humans accustomed to defining identity by labor must recalibrate their sense of purpose. Practices like mindfulness, meditation, and contemplative inquiry help individuals navigate this transition. Emotional resilience and self-compassion are critical tools. Flourishing requires recognizing that value is inherent, not conditional on tasks completed or results achieved.

Education systems must mirror this broader understanding of human value. Schools should cultivate curiosity, moral reasoning, creativity, and relational intelligence. Students must be equipped to navigate a world where answers are immediate but interpretation, judgment, and purpose remain human domains. Teachers become guides in helping young people discover meaning beyond achievement metrics. Learning shifts from accumulation to engagement with ethical, emotional, and creative capacities. Preparing humans for life beyond productivity is as much about character as it is about skill.

Workplaces, too, must adjust to cultivate meaning rather than extract output. Organizations can design roles that emphasize creativity, mentorship, and ethical contribution. Recognition systems should value relational and social impact as much as technical results. Leaders play a pivotal role in fostering environments where humans feel empowered and purposeful. Machines handle tasks efficiently, while humans focus on vision, judgment, and relational influence. Flourishing arises when organizations prioritize human dignity alongside productivity.

The shift beyond productivity invites humans to redefine purpose on their own terms. Tasks and roles are no longer the sole arbiters of meaning; contribution, creativity, and ethical engagement become central. Humans can explore intellectual, relational, and spiritual dimensions that were previously constrained by the demands of work. Autonomy, intention, and moral responsibility become the markers of fulfillment. Flourishing is measured not by efficiency or output but by impact, insight, and connection. This reorientation transforms anxiety about relevance into opportunity for growth and self-discovery.

Humans must also embrace the interplay between technology and ethical agency. AI can perform calculations, analyze trends, and even generate ideas, but it cannot weigh consequences or cultivate empathy. Machines become collaborators, amplifying human potential when guided consciously. Responsibility shifts from merely completing tasks to shaping how intelligence—human and artificial—is applied. Ethical discernment becomes a core competency. Flourishing depends on integrating moral reflection with creative and relational action.

Communities play a critical role in supporting life beyond productivity. Social networks, cultural institutions, and educational systems provide frameworks for purpose and meaning. Collective recognition of value independent of labor reinforces self-worth and fosters resilience. Shared understanding and mutual support mitigate the existential stress of a shifting economy. Flourishing is amplified when humans cultivate networks of care, mentorship, and ethical engagement. Community and collaboration are inseparable from individual growth.

The cultural narrative must evolve to reflect this broader understanding of human value. Society must celebrate the uniquely human capacities that machines cannot replicate—empathy, wisdom, ethical judgment, and creativity. Media, storytelling, and education shape perceptions of worth beyond economic output.

Shifting these narratives helps individuals internalize a sense of purpose grounded in character rather than productivity. Flourishing emerges from a collective recognition of what humans uniquely bring to the world. Identity, dignity, and contribution are no longer tied exclusively to labor.

Finally, embracing humanity beyond productivity is both a personal and societal imperative. Individuals must cultivate awareness, reflection, and ethical judgment. Societies must provide support, recognition, and opportunity beyond work. When humans assert the intrinsic value of being rather than doing, they unlock the potential for deeper fulfillment, creativity, and connection. Flourishing is not the absence of work, but the conscious integration of work, purpose, and human capacity. In this post-productivity era, humans are invited to redefine what it truly means to live a meaningful and valuable life.

Author Reflection:

Writing this chapter forced me to confront my own attachment to productivity as a measure of worth. I realized how deeply ingrained cultural narratives shape identity and self-esteem. By examining what makes life meaningful beyond work, I discovered that value is relational, ethical, and creative—not simply output-based. Recognizing this allows both individuals and societies to cultivate environments where humans thrive independently of labor metrics. Flourishing is possible when worth is understood as intrinsic, not conditional. This perspective transforms anxiety about automation into a framework for human empowerment.

Inspirational Quote:

"*You are not defined by what you produce, but by what you inspire, nurture, and create.*" — Anonymous

Chapter . 6

Creativity After the Algorithm

Human imagination evolves rather than disappears when machines can create.

The arrival of AI capable of generating art, music, and literature has sparked both awe and anxiety. Machines can produce impressive works at scale, often imitating styles of human masters. Some fear that human creativity is no longer unique or valuable. Yet rather than replacing imagination, these technologies redefine it. Creativity becomes a dialogue between human intuition and machine-assisted generation. Humans now explore possibilities beyond their own cognitive limits, expanding the frontier of creative expression.

Machines excel at pattern recognition, recombination, and rapid iteration. They can analyze massive datasets and generate outputs that humans might never imagine alone. This efficiency challenges traditional notions of originality, authorship, and inspiration. Artists and creators must learn to collaborate with machines, treating them as tools that augment rather than replace their vision. The creative process transforms into one of curation, refinement, and ethical guidance. Humans retain the power to choose meaning, context, and purpose.

Creativity after the algorithm emphasizes intention and emotional depth. Machines can mimic style or generate novelty, but they lack lived experience, context, and emotional resonance. Human creators provide purpose, narrative, and ethical grounding. AI amplifies ideas but cannot infuse them with moral or existential significance. The dialogue between human and machine becomes a site of innovation, where imagination is extended rather than substituted. This partnership encourages experimentation and challenges preconceived boundaries.

The democratization of creative tools also changes access to artistic expression. Anyone with a computer or smartphone can now generate complex images, music, or text. While this expands opportunity, it also challenges traditional markers of skill and accomplishment. Humans must cultivate discernment to navigate abundance, distinguishing meaningful work from mere production. Creativity evolves from solitary genius to collaborative engagement, emphasizing interpretation and vision. The focus shifts from quantity to depth, curation, and resonance.

Creativity after the algorithm also invites reflection on the nature of originality. If machines can generate endless permutations of existing ideas, what does it mean to be truly original? Human creativity becomes less about producing novel content in isolation and more about contextual insight, ethical vision, and emotional resonance. Machines expand the palette, but humans determine which combinations carry significance. The focus shifts from invention alone to interpretation, meaning-making, and communication. In this way, AI becomes a collaborator, not a competitor, in the creative journey.

Collaboration with AI allows for experimentation at unprecedented speed. Artists, designers, and writers can test multiple variations instantly, exploring directions that would have taken weeks or months manually. This accelerates the creative cycle and encourages risk-taking. Human judgment and taste remain central in selecting and refining outcomes. The partnership between intuition and computation produces work that neither could achieve alone. Creativity becomes a co-creation, a conversation between human vision and algorithmic suggestion.

Educational approaches to creativity must evolve alongside technological capabilities. Students should learn not only traditional artistic skills but also how to leverage AI tools responsibly. Critical thinking, ethical reasoning, and curatorial judgment become central to the creative curriculum. Learners must understand how to guide machines, interpret outputs, and infuse work with human meaning. Education shifts from teaching technical proficiency alone to fostering holistic creative fluency. Humans develop the capacity to integrate intelligence, intention, and imagination.

Psychological aspects of creativity are also affected by AI. Some may feel threatened, believing that machines diminish their value as creators. Others may feel liberated, using algorithms to extend their imaginative reach. Embracing AI as a collaborator requires openness, flexibility, and humility. Humans must balance trust in machine suggestions with discernment and ethical reflection. Creativity becomes a practice of both inspiration and responsibility. Flourishing arises when humans reclaim imagination as an intentional, human-centered endeavor.

Creativity after the algorithm also transforms professional landscapes. Designers, musicians, and writers can produce more, iterate faster, and explore more complex possibilities than ever before. This changes expectations for output, collaboration, and skill. Success increasingly depends on the ability to guide machines thoughtfully, make aesthetic judgments, and infuse work with human meaning. Technical mastery of AI tools complements traditional artistic expertise. Human imagination remains the driver, while AI acts as an amplifier and accelerator.

Cultural and social implications of machine-assisted creativity are profound. Mass production of content can democratize access to art, literature, and design, but it also risks oversaturation. Distinguishing meaningful work from novelty becomes a human responsibility. Humans determine which ideas resonate culturally, ethically, and emotionally. Curatorial judgment becomes a key creative skill. Flourishing depends on humans acting as stewards of both artistic quality and societal impact.

Ethical considerations in creativity are amplified by AI. Machines can reproduce styles, generate imagery, or mimic voices without consent or understanding of ownership. Human oversight is essential to prevent exploitation, misrepresentation, or cultural appropriation. Ethical guidance becomes inseparable from the creative process. Humans remain accountable for the consequences of machine-assisted work. Creativity is therefore both a technical and moral endeavor, requiring wisdom alongside imagination.

Education must adapt to foster these skills for the next generation. Students need guidance in integrating AI tools with critical thinking, ethical awareness, and personal expression. Training in discernment, judgment, and reflective practice becomes as important as technical skill. Learners must navigate abundance responsibly, curating outputs for meaning rather than volume. The cultivation of human-centered creativity ensures that imagination remains relevant and ethically guided. Flourishing in this environment requires balancing innovation with moral responsibility.

Creativity after the algorithm also reshapes collaboration among humans. Teams can integrate AI-generated ideas as starting points, allowing human collaborators to focus on interpretation, refinement, and ethical judgment. Brainstorming and ideation accelerate, but the role of human insight remains central. Machines supply possibilities; humans supply vision. This shift challenges traditional hierarchies, emphasizing the value of judgment, emotional intelligence, and ethical responsibility. Flourishing arises when teams balance speed and scale with human discernment and relational skill.

The abundance of creative output also changes audience expectations. Consumers encounter unprecedented volume and variety, making curation, context, and meaning more important than sheer novelty. Humans must act as interpreters, guiding audiences toward work that resonates emotionally and ethically. Machines cannot replace human understanding of cultural context or lived experience. Artists, educators, and critics gain prominence as mediators between machine-generated outputs and human experience. Creativity becomes relational, connecting creators, audiences, and ethical frameworks.

Psychological adaptation is necessary for humans to thrive creatively in the AI era. Some may feel threatened, perceiving machine-generated work as competition. Others may experience liberation, using AI to explore ideas previously inaccessible. Mindset becomes as important as skill. Curiosity, resilience, and adaptability are critical for navigating collaboration with intelligent systems. Humans must maintain agency over creative decisions while integrating machine insight responsibly. Flourishing depends on cultivating both confidence and humility.

Ethical awareness shapes both the process and the product of creativity. Humans must consider cultural sensitivity, intellectual property, and the potential societal impact of AI-assisted work. Decisions about what to create, how to deploy it, and the message it communicates carry moral weight. Machines can execute, suggest, and iterate, but cannot bear responsibility. Ethical guidance ensures that human values remain central in creative endeavors. Flourishing is inseparable from conscientious stewardship of both technology and imagination.

As creativity enters a new era, humans are invited to reconsider the very meaning of imagination. Machines can produce endless variations, but they cannot generate purpose, emotional resonance, or ethical insight. Humans define what matters, selecting and refining outputs that align with intention and values. The creative act becomes an exercise in stewardship, integrating intuition, judgment, and moral responsibility. Flourishing arises when humans embrace this expanded role, using AI as a partner rather than a replacement.

Collaboration with machines encourages experimentation and risk-taking. AI can model countless possibilities, freeing humans from constraints of time, skill, or fatigue. This allows creators to explore uncharted directions, test hypotheses, and iterate rapidly. Humans maintain authority over interpretation, style, and meaning. By blending intuition with algorithmic insight, the creative process becomes richer, more agile, and more impactful. Flourishing emerges through co-creation that honors both imagination and conscience.

Social and cultural dimensions of creativity are also transformed. Audiences, communities, and institutions play a role in interpreting and valuing work. Humans mediate between machine outputs and societal relevance, ensuring cultural resonance and ethical integrity. Collaboration becomes multidirectional, involving human creators, machines, and social contexts. Flourishing is defined not just by the production of work, but by its contribution to shared meaning, understanding, and enrichment.

Education and lifelong learning remain essential for navigating AI-assisted creativity. Humans must cultivate critical thinking, ethical awareness, and adaptive skills alongside technical proficiency. Reflective practice, mentorship, and community engagement foster discernment and purpose. AI expands possibilities, but humans must provide vision, coherence, and ethical grounding. Flourishing depends on developing the capacity to integrate intelligence, creativity, and moral responsibility throughout life.

Finally, creativity after the algorithm highlights the enduring uniqueness of human insight. Machines are tools of amplification, but imagination, ethical judgment, and emotional depth remain human domains. The partnership with AI offers unprecedented opportunities while reinforcing the value of distinctly human capacities. Flourishing arises when humans embrace this responsibility, guiding both creation and impact. In a world of algorithmic abundance, creativity becomes a conscious, ethical, and deeply human endeavor.

Author Reflection:

Writing this chapter reminded me that creativity is never solely about output; it is about purpose, meaning, and ethical engagement. AI challenges assumptions about originality but also expands opportunity. I realized that machines do not diminish human imagination—they amplify it when humans remain intentional and discerning. True creative flourishing emerges from collaboration, ethical responsibility, and the courage to explore new possibilities. Human imagination thrives when paired with conscience and curiosity.

Inspirational Quote:

"*The purpose of art is not to compete with machines, but to express the humanity that machines cannot replicate.*" — Anonymous

Chapter . 7

Emotional Intelligence as a Human Frontier

Empathy, care, and emotional presence become more central, not less.

In the age of AI, humans are discovering that emotional intelligence is a critical frontier. Machines may excel at computation, analysis, and pattern recognition, but they cannot truly understand or respond to feelings. Empathy, compassion, and social awareness remain uniquely human skills. As repetitive and cognitive tasks are automated, the human ability to navigate emotions, relationships, and social dynamics becomes even more valuable. Emotional intelligence shapes how individuals collaborate, lead, and build meaningful connections. It is both a professional and deeply personal skill.

Understanding one's own emotions is the foundation of emotional intelligence. Self-awareness allows humans to identify feelings, motivations, and responses in complex situations. Machines cannot experience internal states or reflect on subjective experiences. Humans who cultivate introspection gain clarity, resilience, and authenticity. Emotional regulation enables thoughtful responses rather than reactive behavior. Flourishing requires attention not only to external tasks but to inner life.

Empathy extends beyond individual relationships to collective understanding. Humans interpret the feelings and perspectives of others, navigating diverse experiences and cultural norms. Machines may model human behavior, but they cannot truly inhabit another person's experience. Emotional intelligence allows humans to mediate conflict, foster collaboration, and nurture trust. It is essential for leadership, education, and community-building. Flourishing is relational as well as cognitive, grounded in care and understanding.

Social intelligence is closely linked to professional success and personal well-being. In work and community, the ability to read emotional cues, adapt communication, and respond appropriately is crucial. AI may provide data, forecasts, and suggestions, but it cannot grasp subtleties of tone, context, or intention. Humans who excel at emotional intelligence can bridge gaps between people, cultures, and disciplines. These skills are increasingly the differentiator in an automated world. Human value is amplified through connection and responsiveness.

Emotional intelligence also plays a pivotal role in leadership. Leaders who understand and manage emotions effectively inspire trust, loyalty, and engagement. Machines can provide data-driven recommendations, but they cannot motivate, encourage, or empathize with teams. Emotional awareness allows humans to navigate conflicts, recognize strengths, and foster a culture of inclusion and collaboration. Leadership rooted in empathy cultivates both productivity and well-being. Flourishing organizations prioritize emotional intelligence as a core competency, not a soft skill.

In education, emotional intelligence shapes how students learn and interact. Teachers who model empathy, emotional awareness, and social sensitivity create environments where curiosity and creativity flourish. Machines can provide personalized instruction, but they cannot nurture the subtleties of human growth. Students benefit from guidance, mentorship, and relational support that fosters resilience and self-confidence. Emotional intelligence equips learners to navigate both technological and social challenges. Flourishing emerges when humans develop both cognitive and emotional literacy.

Cultural intelligence is deeply connected to emotional intelligence. Understanding and responding appropriately to diverse cultural norms and emotional expressions is critical in an interconnected world. AI can simulate cultural patterns, but genuine sensitivity requires lived experience and ethical judgment. Humans with strong emotional intelligence can bridge divides, mediate misunderstandings, and foster collaboration across differences. These skills are essential for global teamwork, diplomacy, and social cohesion. Flourishing depends on the ability to connect across boundaries with awareness and respect.

Emotional intelligence also supports ethical decision-making. Recognizing how actions affect others, anticipating emotional consequences, and responding with care are critical aspects of moral responsibility. Machines may optimize efficiency or outcomes, but they cannot evaluate the emotional or ethical dimensions of decisions. Humans with high emotional intelligence integrate empathy with reasoning, ensuring that choices are compassionate and just. Flourishing societies rely on these capacities to balance innovation with human welfare. Emotional intelligence safeguards both humanity and ethics in a rapidly changing world.

Emotional intelligence also enhances creativity and innovation. Humans who are attuned to emotions can channel feelings, motivations, and insights into original ideas and solutions. Machines can generate content or suggest possibilities, but they cannot experience passion, curiosity, or inspiration. Human creativity is enriched by emotional awareness, allowing for depth, authenticity, and resonance. Collaboration between humans and AI is most effective when guided by emotional insight. Flourishing emerges when emotion informs imagination, giving ideas meaning and impact.

In the workplace, emotional intelligence is a key differentiator. Teams composed of technically skilled members may still struggle without empathy, communication, and interpersonal understanding. Humans who can navigate conflict, motivate colleagues, and interpret nuanced social cues foster cohesion and productivity. AI tools may enhance efficiency, but they cannot replace trust, rapport, or ethical judgment. Organizations that prioritize emotional intelligence cultivate cultures that are resilient, innovative, and ethically sound. Flourishing requires balancing technological capability with human relational skill.

Emotional intelligence also strengthens personal relationships. Families, friendships, and communities rely on empathy, active listening, and emotional support. Machines can simulate interactions or provide companionship, but they cannot authentically respond to human feelings or navigate relational complexity. Humans capable of emotional attunement create deeper connections, mutual understanding, and social cohesion. Flourishing individuals contribute not only through work or productivity but through care, presence, and connection with others. Emotional intelligence is thus foundational to both personal and societal well-being.

Education systems must nurture emotional literacy alongside academic skills. Developing self-awareness, empathy, and social competence prepares students for collaborative work, global interaction, and ethical decision-making. AI may assist in delivering content, but it cannot model or teach emotional nuance. Human educators guide students in recognizing feelings, understanding perspectives, and managing responses. Emotional intelligence becomes as important as technical skill in preparing the next generation for a world dominated by both machines and humans. Flourishing requires emotional as well as cognitive development.

Emotional intelligence becomes increasingly vital as society becomes more automated and technologically complex. Humans must navigate work environments, communities, and social networks that are mediated by machines. AI can facilitate communication and collaboration, but it cannot interpret subtle emotional cues or resolve interpersonal tension. Humans with strong emotional intelligence act as bridges, translating, mediating, and fostering understanding across individuals and groups. Social cohesion, trust, and collaboration are maintained through empathy and insight. Flourishing requires a conscious focus on human connection amid technological complexity.

Mental health and well-being are deeply tied to emotional intelligence. Awareness of one's own emotions and the emotions of others allows humans to manage stress, prevent burnout, and sustain meaningful engagement. Machines may provide analytics, recommendations, or mental health support tools, but the human capacity for empathy, listening, and emotional support remains irreplaceable. Emotional intelligence fosters resilience, perspective, and self-compassion. In a world of constant change and automation, these qualities are essential to thriving, not just surviving.

Emotional intelligence also drives ethical action in complex, technology-driven environments. Humans must evaluate the emotional impact of decisions on colleagues, clients, communities, and society at large. Machines cannot weigh consequences that require empathy or moral judgment. Individuals who cultivate emotional intelligence are better equipped to guide AI and organizational practices responsibly. They anticipate harm, mediate conflicts, and ensure fairness. Flourishing requires humans to wield intelligence with both ethical awareness and emotional sensitivity.

The capacity to connect emotionally also supports learning and growth. Humans who can engage empathetically with diverse perspectives, feedback, and experiences expand their understanding and adaptability. Emotional intelligence enhances collaboration, mentoring, and coaching, creating environments that foster curiosity and innovation. Machines cannot replicate this relational dimension, making human emotional insight a central driver of development. Flourishing is therefore relational as well as cognitive, grounded in ethical, empathetic engagement.

Emotional intelligence ultimately redefines human relevance in an automated world. As machines handle tasks requiring speed, precision, and pattern recognition, humans are called to exercise empathy, ethical judgment, and relational insight. The ability to understand and respond to emotions becomes a primary source of influence and impact. Flourishing depends not on what humans can compute, but on what they can feel, understand, and communicate. Emotional intelligence transforms challenges into opportunities for connection, collaboration, and ethical action.

In leadership, emotional intelligence ensures that teams remain cohesive, motivated, and resilient. Leaders who can empathize, listen, and navigate conflict inspire loyalty and trust. AI can provide data, forecasts, and recommendations, but humans supply nuance, compassion, and ethical perspective. Organizations that prioritize emotional intelligence foster environments where creativity, productivity, and well-being coexist. Flourishing emerges when human insight directs both interpersonal and technological systems.

In personal life, emotional intelligence strengthens bonds with family, friends, and community. Recognizing emotions, practicing active listening, and responding with care cultivates trust and intimacy. Machines can assist communication but cannot provide authentic emotional reciprocity. Humans who invest in relational awareness contribute to networks of support, understanding, and shared meaning. Flourishing depends on these connections as much as on individual achievement or technological mastery.

Education and professional development must integrate emotional intelligence intentionally. Students and employees alike benefit from learning to interpret emotions, navigate social dynamics, and engage ethically. Emotional literacy complements technical proficiency, preparing humans to thrive alongside AI. Flourishing arises when education cultivates both mind and heart, enabling humans to guide technology and society with empathy and insight. Emotional intelligence is a lifelong skill that enriches every domain of human experience.

Chapter . 8

Relationships in a Mediated Reality

AI reshapes how humans connect, love, and form community.

The digital age has already transformed human relationships, but artificial intelligence accelerates this shift in profound ways. Social media algorithms, chatbots, and recommendation engines mediate interactions, influencing who we connect with and how. AI can suggest friends, partners, or collaborators, while shaping the content we see and respond to. These tools offer unprecedented opportunities for connection, but they also introduce complexity, bias, and dependency. Humans must learn to navigate relationships in a world where interactions are increasingly filtered, guided, or augmented by technology. Flourishing requires intentionality in both digital and physical connections.

AI-driven communication tools can enhance efficiency, intimacy, and accessibility. People separated by distance can maintain meaningful relationships through virtual assistants, translation tools, and personalized interaction platforms. Machines can help schedule meetings, manage correspondence, or even suggest empathetic responses. These tools augment human capability, freeing time and mental energy for deeper engagement. Yet reliance on AI also risks diluting authenticity, reducing the richness of unmediated emotional expression. The challenge is balancing technological convenience with genuine presence.

Digital mediation also raises questions of trust and authenticity. AI can simulate human interaction, generating responses that appear empathetic or conversational. Distinguishing genuine human engagement from algorithmically produced communication becomes increasingly important. Relationships thrive on trust, reciprocity, and shared understanding—qualities that cannot be fully replicated by machines. Humans must cultivate discernment, emotional awareness, and ethical boundaries to navigate mediated social spaces. Flourishing depends on preserving authenticity amid automation.

Intimacy and attachment are reshaped by AI's presence. Virtual companions, AI-driven social platforms, and predictive matchmaking algorithms influence romantic and familial bonds. Machines can suggest partners, provide conversational support, or simulate companionship, but they cannot truly experience love, loyalty, or care. Humans must take responsibility for emotional well-being, consent, and ethical interaction. Flourishing relationships require balancing technological facilitation with personal accountability and empathy. AI becomes a tool for connection, not a substitute for human presence.

Mediated relationships also affect emotional development. Humans increasingly experience social cues through screens, avatars, and AI filters. Subtle aspects of communication—tone, body language, timing—can be distorted or lost. Emotional literacy requires deliberate effort, practicing empathy and active listening even in digital spaces. Machines can provide feedback or suggest improvements, but humans remain responsible for meaningful engagement. Developing relational skills becomes both more challenging and more essential in a mediated reality. Flourishing depends on maintaining emotional presence despite technological intermediaries.

The interplay between AI and social norms is profound. Algorithms can reinforce certain behaviors by promoting content that aligns with engagement metrics rather than ethical or social considerations. Individuals may unconsciously adapt to these mediated cues, shaping attitudes, expectations, and behavior.

Humans must critically reflect on how technology influences interaction patterns and societal values. Conscious awareness enables ethical, intentional, and empathetic participation. Flourishing requires humans to assert agency in shaping social norms rather than passively following algorithmic patterns.

Mediated reality also impacts trust and conflict resolution. Communication through AI-assisted channels may reduce face-to-face interactions, complicating accountability and emotional clarity. Misunderstandings can escalate when context or nuance is lost in digital translation. Humans must develop conflict management strategies, ethical reasoning, and relational intelligence tailored to mediated contexts. Machines can provide analysis or suggestions, but emotional and ethical judgment remains irreducibly human. Flourishing communities balance technological efficiency with relational integrity.

AI-mediated relationships also challenge identity formation and self-expression. Online personas, avatars, and curated content create layers of representation that can differ from offline reality. Individuals may struggle to reconcile these versions of self, leading to identity confusion or social pressure. Humans must navigate authenticity, self-awareness, and ethical presentation in both digital and physical spheres. Emotional intelligence, ethical reflection, and relational insight are critical tools. Flourishing requires coherence between digital and human selves, ensuring meaningful and ethical connections.

AI-mediated communication also transforms intimacy and romantic relationships. Dating apps, recommendation algorithms, and virtual companionship platforms can suggest matches based on data patterns and behavioral predictions. While these tools increase access and efficiency, they cannot replicate human intuition, emotional depth, or mutual growth. Humans must navigate ethical boundaries, consent, and emotional honesty to ensure meaningful connections. Flourishing romantic relationships require intentionality, empathy, and reflective engagement beyond algorithmic suggestions. Technology can facilitate encounters, but humans remain the ultimate authors of connection and care.

Friendship and peer networks are similarly reshaped. AI can suggest social groups, recommend activities, or even mediate conversations in ways that expand opportunities for connection. Yet the quality of friendship depends on trust, vulnerability, and shared experience—dimensions machines cannot authentically generate. Human agency is required to cultivate empathy, presence, and reciprocity. Emotional intelligence and ethical consideration remain central to sustaining supportive and meaningful peer networks. Flourishing social bonds are guided by humans even as technology amplifies reach and accessibility.

Intergenerational relationships face unique challenges in a mediated reality. Older generations may struggle with technological literacy, while younger generations grow up immersed in AI-assisted interactions. These differences can create gaps in understanding, emotional expression, and trust. Humans must intentionally bridge generational divides, fostering patience, mentorship, and guidance. Emotional intelligence becomes critical for translating experiences and values across age groups. Flourishing occurs when mediated technology strengthens, rather than weakens, intergenerational connection.

Professional relationships are also affected by AI mediation. Teams increasingly collaborate across geographies, time zones, and cultures with the assistance of AI tools. Communication, task management, and project coordination are optimized by algorithms, but relational nuance requires human oversight. Trust, ethical judgment, and cultural sensitivity guide effective collaboration. Flourishing workplaces balance technological efficiency with human relational skill, fostering engagement, inclusion, and ethical decision-making. AI can facilitate processes, but humans define the quality of interaction.

Mediated reality also reshapes how humans experience empathy and moral responsibility. When interactions occur through screens or AI intermediaries, the emotional weight of others' experiences can feel abstract or distant. Humans must actively cultivate sensitivity to the feelings, struggles, and perspectives of those they interact with digitally. Ethical reasoning and relational mindfulness are essential for ensuring that mediated connections do not erode compassion. Machines can provide reminders or prompts, but genuine empathy requires human reflection and care. Flourishing depends on bridging the gap between digital mediation and authentic emotional engagement.

The rise of AI companions and synthetic relationships introduces new ethical and psychological dimensions. Chatbots, virtual mentors, and AI friends can provide support, feedback, and companionship, but they cannot experience reciprocity or moral understanding. Humans interacting with synthetic agents must remain aware of boundaries, emotional authenticity, and the potential for dependency. Flourishing relationships require maintaining human-to-human connection as the foundation, with AI acting as an adjunct, not a replacement. Emotional and ethical guidance is necessary to navigate these new forms of interaction responsibly.

Mediated reality also amplifies the risk of echo chambers and social polarization. AI-driven algorithms often prioritize engagement over accuracy, promoting content that reinforces preexisting beliefs. Humans must exercise discernment, critical thinking, and ethical judgment to avoid isolation in biased networks. Cultivating awareness, empathy, and dialogue across perspectives is essential for social cohesion. Flourishing in a mediated reality requires actively resisting the narrowing of perception and fostering inclusive, diverse, and respectful communication. Technology must be guided by human values, not left to shape behavior passively.

Education and mentorship play critical roles in fostering relational intelligence in a mediated world. Teaching young people to navigate digital spaces with empathy, discernment, and ethical awareness prepares them for both personal and professional life. Human guidance ensures that AI tools are used responsibly, relationships are nurtured thoughtfully, and social consequences are considered. Flourishing emerges when humans are equipped to integrate technological assistance with emotional and ethical competence. Relationships remain central, even as the tools mediating them evolve.

As mediated reality becomes the norm, humans are called to redefine intimacy, trust, and community. Relationships are no longer confined to physical presence but span digital spaces, networks, and algorithmically suggested connections. AI can facilitate introductions, support dialogue, and manage interactions, but it cannot provide the moral, emotional, or ethical foundation necessary for true human connection. Humans must remain active participants in shaping relational quality, ensuring authenticity, reciprocity, and care. Flourishing relationships are intentional, guided by empathy and ethical reflection, not automated convenience.

The ethical dimension of mediated relationships is paramount. Humans must evaluate the consequences of technology on emotional well-being, social cohesion, and moral accountability. Algorithms may suggest interactions or filter information, but humans decide how to respond, engage, and nurture trust. Awareness of bias, digital influence, and potential harm is essential. Emotional intelligence and ethical reasoning become central to navigating the opportunities and risks of mediated reality. Flourishing is achieved when human values guide both technology and interaction.

Mediated reality also challenges notions of commitment and responsibility. Digital spaces allow for fluid, rapid connections, but they can undermine relational depth and accountability if not approached consciously. Humans must cultivate patience, attentiveness, and moral engagement to sustain meaningful bonds. Machines provide scale and convenience, but they cannot substitute for relational effort, ethical judgment, or emotional investment. Flourishing arises when humans integrate technology into relationships without relinquishing responsibility for care, empathy, or trust.

Community building in a mediated reality requires intentional human stewardship. Online groups, virtual organizations, and global networks offer immense potential for collaboration, learning, and social change. AI can optimize logistics and suggest synergies, but humans provide the vision, values, and ethical framework. Relationships, trust, and shared purpose are cultivated through active participation, guidance, and reflection. Flourishing communities balance technological assistance with conscious human effort, fostering inclusion, ethical engagement, and meaningful interaction.

Chapter . 9

Spirituality in an Engineered World

AI forces ancient questions about meaning, soul, and transcendence back into focus.

The rise of AI is not just a technological or economic phenomenon; it is also a spiritual challenge. As machines increasingly perform tasks that once defined human intelligence, people are prompted to ask profound questions about purpose, meaning, and existence. What does it mean to be human when computation, pattern recognition, and creative expression can be automated? Spiritual inquiry—questions of morality, transcendence, and the soul—becomes central to human flourishing. Technology cannot answer questions of ultimate meaning, but it can provoke reflection on why such questions matter. Flourishing depends on engaging with these timeless questions in the context of a rapidly evolving world.

AI invites reflection on the nature of consciousness. Machines process information, generate outputs, and simulate intelligence, yet they lack awareness, experience, and intentionality. Humans are confronted with what it means to possess subjective experience and moral responsibility. Spirituality intersects with technology when individuals consider the ethical implications of creation, autonomy, and stewardship. Machines may enhance efficiency or knowledge, but humans remain uniquely capable of seeking truth, purpose, and connection with the transcendent. Flourishing arises from recognizing this distinction.

Automation challenges traditional markers of significance. Many have historically derived meaning from labor, achievement, and contribution. As AI assumes more functions, humans must find purpose beyond productivity. Spiritual exploration offers a framework for understanding value, purpose, and fulfillment in non-material terms. Practices such as meditation, reflection, community engagement, and ethical inquiry help individuals navigate these questions. Flourishing involves cultivating awareness of what truly matters, beyond measurable output or efficiency.

The integration of AI into daily life also raises ethical and existential dilemmas. Decisions about AI design, deployment, and oversight carry moral weight and social consequence. Humans are called to exercise discernment, responsibility, and foresight. Spiritual inquiry offers guidance in evaluating not just what can be done, but what should be done. Flourishing depends on cultivating wisdom alongside technological capability, ensuring that innovation aligns with human values. Ethics and spirituality are intertwined in the pursuit of meaningful action.

AI also compels humans to reconsider the relationship between knowledge and wisdom. Machines can store and process vast amounts of information, generate insights, and even provide recommendations for action. Yet knowledge alone does not answer questions of meaning, morality, or purpose. Spirituality emphasizes discernment, ethical judgment, and reflective awareness—capacities that remain uniquely human. Flourishing depends on integrating intelligence with insight, ensuring that knowledge serves wisdom rather than replacing it. In an engineered world, humans must cultivate the ability to ask why, not just how.

Religious and philosophical traditions offer frameworks for engaging with AI ethically and meaningfully. Concepts such as compassion, stewardship, and the inherent dignity of life guide human interaction with technology. These traditions remind humans that value is not solely measured in efficiency, productivity, or innovation. Machines can optimize outcomes, but humans must interpret meaning, navigate moral complexity, and act with integrity. Flourishing requires drawing on spiritual and ethical wisdom to direct technological power toward human good.

AI also challenges humans to confront mortality and impermanence in new ways. Machines may extend life, enhance health, or preserve information across generations, yet they do not experience existence, suffering, or loss. Spiritual inquiry encourages reflection on what it means to live a meaningful life within finite human limits. Technology can augment capabilities, but humans must define purpose, values, and legacy. Flourishing emerges when individuals integrate technological progress with awareness of life's ethical and existential dimensions.

The contemplative dimension of spirituality becomes more vital in a fast-paced, algorithm-driven world. Humans must cultivate stillness, reflection, and mindfulness to maintain connection with inner values and purpose. AI accelerates work, information, and communication, but it cannot provide inner clarity or moral insight. Flourishing requires intentional practices that nurture ethical discernment, emotional balance, and a sense of transcendence. Spirituality offers tools to navigate complexity, uncertainty, and the moral weight of innovation.

AI also reshapes communal spiritual life. Virtual congregations, online meditation groups, and AI-assisted rituals allow humans to connect across distances and time zones. These tools can enhance accessibility, engagement, and shared practice, yet they cannot replicate the embodied, relational aspects of human gathering. Humans remain responsible for fostering community, ethical interaction, and authentic spiritual experience. Flourishing arises when technology supports, rather than replaces, the human dimensions of shared meaning, ritual, and moral reflection.

The tension between automation and human purpose highlights the need for ethical leadership. Developers, policymakers, and leaders must consider the spiritual and moral implications of technological choices. Decisions regarding AI deployment affect human well-being, social cohesion, and ethical norms. Spiritual reflection provides a framework for evaluating consequences, prioritizing human dignity, and guiding innovation responsibly. Flourishing societies integrate technological capability with moral and ethical deliberation, ensuring that progress serves humanity rather than undermining it.

AI also prompts reflection on humility and interdependence. Machines may surpass humans in speed, memory, or pattern recognition, yet they remain dependent on human intention, guidance, and ethical oversight. Spirituality emphasizes the limits of human and machine knowledge, fostering humility, reflection, and moral responsibility. Recognizing interdependence—between humans, machines, and the broader environment—encourages collaboration grounded in care and ethical stewardship. Flourishing depends on balancing power with responsibility and intelligence with moral awareness.

Personal spiritual practice becomes a crucial anchor amid rapid technological change. Meditation, prayer, reflection, and ethical contemplation help individuals maintain clarity, purpose, and resilience. Machines can support routines, track habits, or provide information, but they cannot cultivate meaning, moral insight, or inner peace. Humans must actively engage in practices that nurture ethical awareness, emotional balance, and purpose. Flourishing emerges when technology is harnessed to deepen, rather than distract from, spiritual and moral life.

The interplay between AI and spirituality also raises questions about moral responsibility. As machines influence decisions, behavior, and social outcomes, humans must actively guide their use with ethical discernment. Spirituality provides a lens for evaluating the consequences of technological action on individuals, communities, and the planet. Flourishing requires integrating moral reflection with technical innovation, ensuring that progress aligns with values of compassion, justice, and human dignity. AI becomes a tool for enacting ethical vision, not a substitute for conscience.

Technology can also deepen human self-understanding. AI-driven analytics and data visualization may reveal patterns in behavior, thought, and emotion that were previously invisible. These insights can support self-reflection, personal growth, and ethical decision-making. Yet machines cannot interpret meaning, prioritize values, or provide moral guidance. Spirituality bridges this gap, guiding humans to translate data into wisdom, insight, and purposeful action. Flourishing arises when humans combine analytical clarity with ethical and existential awareness.

Global interconnectedness amplifies the spiritual dimensions of AI. Algorithms facilitate communication, collaboration, and cultural exchange across borders, but they also expose inequalities, biases, and systemic injustices. Humans must navigate these challenges with moral insight, empathy, and ethical action. Spiritual frameworks provide tools for understanding suffering, promoting equity, and fostering shared meaning. Flourishing requires humans to engage with technology consciously, ensuring that innovation enhances collective well-being rather than perpetuating harm.

The acceleration of life brought by AI intensifies the need for reflection and contemplation. Machines optimize, automate, and accelerate processes, creating pressure to constantly perform and produce. Spiritual practices help humans maintain balance, perspective, and intentionality. Mindfulness, ethical contemplation, and reflective inquiry provide anchors amid rapid technological change. Flourishing is realized when humans resist the tyranny of efficiency and cultivate a life guided by purpose, ethics, and self-awareness. Technology should serve life, not control it.

AI ultimately forces humanity to confront fundamental questions about purpose, meaning, and ethical responsibility. As machines take on tasks once reserved for humans, people are prompted to ask: what is truly essential about being human? Spirituality provides a framework for exploring these questions, offering guidance on morality, connection, and transcendence. Flourishing depends not on technological capability alone, but on human insight, ethical judgment, and the pursuit of meaning. Machines can enhance life, but they cannot provide purpose—this remains the domain of human conscience and reflection.

Ethical stewardship becomes central in a world shaped by AI. Humans must consider the consequences of their creations, weighing benefits against potential harm. Spirituality emphasizes moral responsibility, care for others, and accountability for collective outcomes. Flourishing societies integrate technological progress with ethical reflection, ensuring that machines amplify human well-being rather than undermining it. The choices humans make in designing and deploying AI are as much spiritual as technical, shaping the moral and relational landscape of the future.

Personal reflection and inner cultivation are vital for navigating an engineered world. Practices such as meditation, journaling, prayer, or contemplative inquiry allow individuals to maintain clarity, purpose, and ethical focus. Machines can assist with information and analysis, but they cannot cultivate self-awareness, moral discernment, or emotional resilience. Flourishing requires humans to take responsibility for inner life, integrating reflection with action in both personal and societal spheres. Spiritual engagement anchors human identity in the midst of rapid technological transformation.

Communal spiritual life remains essential, even in AI-mediated environments. Humans form meaning through shared rituals, ethical discourse, and collective pursuit of purpose. Technology can enhance accessibility and connection, but it cannot replace embodied human presence, dialogue, or ethical accountability. Flourishing communities rely on humans to cultivate empathy, shared values, and moral cohesion. AI serves as a tool to support these processes, not a substitute for relational and ethical engagement.

Finally, AI is a mirror reflecting humanity's deepest questions and potentials. Machines extend capabilities, reveal patterns, and simulate intelligence, yet they cannot replace conscience, empathy, or moral insight. Flourishing emerges when humans use technology to deepen understanding, enhance ethical action, and cultivate purpose. Spirituality guides this integration, ensuring that human identity, values, and meaning remain central in a world increasingly shaped by AI. The engineered world becomes a canvas for moral, ethical, and spiritual human flourishing.

Author Reflection:

Writing this chapter reminded me that the rise of AI is not just a technological revolution—it is a spiritual and ethical one. Machines challenge humans to reconsider purpose, morality, and meaning, but they cannot answer these questions for us. I realized that flourishing depends on integrating reflection, ethical responsibility, and relational care into every aspect of life. Spirituality becomes a compass, guiding humans to use AI in ways that enhance well-being, ethical awareness, and connection. Technology amplifies human capacity, but the soul remains distinctly human.

Inspirational Quote:

"*Machines can calculate, create, and optimize—but only humans can seek meaning, embody ethics, and cultivate the soul.*" — Anonymous

Part III — Human-Machine Collaboration

Chapter . 10

The Myth of Man Versus Machine

The future favors partnership over competition between humans and AI.

The narrative of humans versus machines has long captured imaginations, from science fiction to media headlines. Stories of AI overthrowing human labor, intelligence, or creativity dominate public discourse. Yet this framing is misleading and counterproductive. The real opportunity lies in collaboration—humans and machines working together to amplify intelligence, creativity, and problem-solving. Competition is a false dichotomy; flourishing arises from partnership. Humans retain unique capacities for judgment, empathy, and moral insight, while machines provide scale, speed, and pattern recognition. Recognizing this complementarity reframes fear into potential.

Machines excel at processing data, identifying patterns, and performing repetitive or complex computations. Humans excel at interpretation, ethical reflection, creativity, and emotional understanding. By combining these strengths, humans and AI can solve problems that neither could address alone. Collaboration requires intentional design, trust, and clear understanding of roles. The myth of rivalry distracts from the ethical, practical, and relational work required to integrate AI responsibly. Flourishing emerges when humans leverage technology to extend—not replace—their unique capabilities.

Education and professional development must embrace the collaborative model. Humans must learn how to work alongside AI, understanding its capabilities, limitations, and potential biases. Skills such as judgment, ethical reasoning, creativity, and emotional intelligence become central. AI literacy is essential—not to compete with machines, but to collaborate effectively. Teams and organizations that foster this integration are better equipped to innovate, adapt, and thrive. Flourishing arises from cultivating co-dependence, not conflict, between human insight and machine intelligence.

The fear of obsolescence drives the myth of man versus machine. Jobs, decision-making authority, and societal influence are perceived as threatened by intelligent systems. Yet this fear often obscures opportunities for enrichment, augmentation, and new forms of work. Humans who embrace AI as a partner gain capacity, reach, and insight beyond what was previously possible. Technology amplifies human problem-solving rather than diminishing it. Flourishing depends on embracing this mindset, shifting from anxiety to agency in a collaborative future.

Collaboration with AI also reshapes problem-solving approaches. Humans provide context, meaning, and ethical oversight, while machines process information, model scenarios, and generate insights. This complementary interaction allows for faster, more informed, and creative solutions. Teams that embrace co-creation can tackle challenges in science, medicine, policy, and the arts with unprecedented efficiency and depth. The myth of rivalry diminishes when humans recognize that machines are instruments of enhancement rather than threats to relevance. Flourishing emerges from this synergy, where each partner contributes distinct strengths.

Trust is central to effective human–machine collaboration. Humans must understand AI's capabilities, limitations, and decision-making processes to use it effectively. Blind reliance can lead to errors, while excessive skepticism can hinder potential. Transparency in algorithms, explainable AI, and human oversight ensure that collaboration remains ethical and productive. Trust is cultivated not by assuming machines are sentient, but by understanding how they operate and integrating that knowledge responsibly. Flourishing depends on building confidence in partnership through understanding and ethical stewardship.

Collaboration also requires rethinking the role of humans in work and society. As AI assumes routine or analytical tasks, humans are freed to focus on interpretation, strategy, creativity, and relational dimensions. Work becomes less about repetition and more about insight, judgment, and ethical decision-making. This shift demands skill development, lifelong learning, and adaptability. The myth of obsolescence fades when humans recognize their irreplaceable contributions in areas where machines cannot replicate human insight, empathy, or moral reasoning. Flourishing arises from embracing unique human capacities while leveraging machine efficiency.

Ethical reflection is inseparable from collaboration. Humans determine the purposes, boundaries, and consequences of AI deployment. Machines may suggest courses of action, predict outcomes, or generate options, but they cannot weigh moral implications. Flourishing requires humans to lead with ethical foresight, ensuring that technological augmentation aligns with human well-being. Collaboration is not just functional—it is moral, relational, and intentional. Humans maintain responsibility for outcomes, shaping AI as a partner for ethical, meaningful action.

Human–machine collaboration also redefines leadership. Leaders who understand the capabilities and limitations of AI can guide teams more effectively, balancing human judgment with algorithmic support. Decision-making becomes more data-informed, but human insight, ethical reasoning, and strategic vision remain central. Machines provide analysis and scenario modeling, but they cannot provide purpose, moral clarity, or empathy. Leadership in the AI era is a partnership—human values steering technological power toward collective benefit. Flourishing emerges when leaders cultivate both technical literacy and relational intelligence.

Education and workforce development play critical roles in fostering effective collaboration. Learning how to communicate with, interpret, and guide AI systems becomes as essential as traditional skills. Students and employees must develop adaptability, critical thinking, ethical judgment, and emotional intelligence to navigate complex, AI-mediated environments. Technical proficiency alone is insufficient; humans must also be adept at leveraging AI responsibly and creatively. Flourishing arises when people are prepared not to compete with machines, but to enhance their own capacities through collaboration.

Collaboration also amplifies creativity and innovation. AI can generate possibilities, model alternative solutions, and uncover patterns that humans might miss. Yet interpretation, meaning-making, and ethical judgment remain human responsibilities. Artists, designers, scientists, and thinkers who integrate AI into their creative processes can achieve outcomes previously unimaginable. Machines accelerate experimentation, but humans provide intentionality and cultural resonance. Flourishing emerges from co-creation, where technology extends human imagination rather than replacing it.

Social and cultural collaboration with AI requires sensitivity to equity and inclusion. Access to AI tools, bias in algorithms, and unequal distribution of technological resources can reinforce social disparities. Humans must ensure that collaboration is guided by ethical principles, promoting fairness, justice, and opportunity. Partnering with AI responsibly demands awareness of societal impact and deliberate effort to prevent harm. Flourishing communities are those where human–machine partnerships are aligned with shared values and collective well-being.

Human–machine collaboration also transforms problem-solving on a societal scale. AI can analyze massive datasets, simulate complex systems, and forecast outcomes, enabling humans to make more informed, strategic decisions. Yet the interpretation of data, the weighing of ethical implications, and the prioritization of societal values remain human responsibilities. Collaboration ensures that technological power is directed by moral and social considerations, rather than left to the mechanical logic of algorithms alone. Flourishing communities emerge when humans and AI work together to address challenges that neither could tackle independently.

In research and innovation, collaboration accelerates discovery. AI can process literature, identify patterns, and suggest hypotheses, providing scientists and innovators with unprecedented tools. Humans, however, bring intuition, ethical judgment, and domain expertise to determine which directions are meaningful and responsible. This partnership fosters breakthroughs in medicine, sustainability, and technology that align with human values. The myth of rivalry fades as humans recognize that machines enhance, rather than threaten, intellectual and creative work. Flourishing arises when collaboration is intentional, ethical, and aligned with purpose.

Collaboration requires cultivating trust between humans and AI. Users must understand the limitations, biases, and decision-making processes of intelligent systems. Transparent algorithms, explainable AI, and human oversight are essential for responsible integration. Trust allows humans to rely on AI outputs while retaining critical judgment, ensuring that ethical and social considerations remain central. Flourishing depends on balancing confidence in AI with human discernment, fostering both competence and accountability in collaborative endeavors.

Human relationships and social dynamics are also influenced by collaboration with AI. Teams can work more efficiently across distances, time zones, and cultures with the assistance of AI tools. Yet relational intelligence, empathy, and ethical negotiation remain human responsibilities. Machines facilitate communication and information flow but cannot replace the moral and emotional labor required to build trust and cohesion. Flourishing in collaborative environments emerges from human stewardship, relational skill, and ethical guidance, supported—not supplanted—by technology.

The partnership between humans and machines also invites reflection on responsibility and stewardship. Humans determine the purpose, scope, and boundaries of AI deployment, shaping its impact on society, ethics, and culture. Machines may execute tasks efficiently, but humans provide moral vision, contextual judgment, and long-term oversight. Flourishing depends on recognizing this responsibility, ensuring that AI serves human well-being, equity, and creativity. Collaboration is thus both a practical and ethical imperative.

Intentional integration of AI into human work and life requires flexibility, adaptability, and continuous learning. Humans must cultivate skills that complement machine capabilities, including ethical reasoning, emotional intelligence, and critical thinking. Teams that embrace these principles can innovate, solve complex problems, and enhance collective potential. The myth of rivalry dissipates when collaboration is approached with clarity, purpose, and mutual respect. Flourishing is achieved when humans and machines operate as partners rather than competitors.

Creativity, insight, and ethical judgment remain uniquely human domains, even as AI expands computational and analytical reach. Humans provide interpretation, meaning, and values that machines cannot replicate. Collaborative systems amplify human ingenuity, allowing for new forms of discovery, art, and societal progress. Flourishing emerges when humans harness AI as a tool for innovation while retaining control over the ethical and relational dimensions of work. Partnership, not replacement, defines the future of human relevance.

Human–machine collaboration also strengthens resilience and problem-solving capacity. In complex global challenges—climate change, public health, inequality—AI can process information at scale, while humans provide judgment, ethical prioritization, and long-term perspective. Together, humans and machines create solutions that neither could achieve alone. Flourishing communities emerge when technological power is guided by human wisdom, empathy, and ethical vision.

Finally, collaboration transforms our understanding of what it means to be human. The myth of man versus machine distracts from the true potential: amplifying human creativity, judgment, and moral reasoning through partnership with AI. Humans remain central as curators, ethical stewards, and interpreters of meaning, while machines extend reach and capacity. Flourishing arises when humans embrace collaboration as a pathway to growth, impact, and purpose in an increasingly intelligent world.

Author Reflection:

Writing this chapter emphasized that fear of machines often overshadows opportunity. Humans are not in competition with AI; we are in a partnership with it. I realized that flourishing depends on intentional collaboration, ethical stewardship, and recognition of our unique strengths. Machines amplify human potential but cannot replace moral insight, empathy, or creativity. True progress emerges when we embrace AI as a partner, guiding it to serve human well-being and collective flourishing.

Inspirational Quote:

"The future is not man versus machine—it is humans guided by wisdom, partnering with intelligence to create a better world." — Anonymous

Chapter . 11

Augmented Intelligence, Not Artificial Replacement

AI amplifies human capability when designed with intention.

Page One

The notion of augmented intelligence reframes the relationship between humans and machines. Rather than replacing humans, AI can enhance their cognitive, creative, and decision-making capacities. Machines process data at scale, identify patterns, and suggest solutions, while humans provide context, judgment, and ethical guidance. This collaboration transforms work, learning, and innovation, emphasizing enhancement over obsolescence. Augmentation requires intentional design, ensuring that technology complements rather than diminishes human capability. Flourishing arises when humans leverage AI to extend their unique strengths.

Augmented intelligence differs fundamentally from automation. Automation replaces repetitive, rule-based tasks, reducing human involvement in predictable processes. Augmentation, on the other hand, positions humans at the center of decision-making, supported by AI insights and analysis. Professionals in medicine, engineering, research, and creative fields increasingly rely on AI as a partner to improve accuracy, efficiency, and insight. Humans retain control over interpretation, ethical evaluation, and strategic choice. Flourishing is achieved when technology amplifies human potential rather than supplanting it.

The design of AI systems is critical to effective augmentation. Systems must be transparent, explainable, and aligned with human values to support collaboration. When humans understand how AI arrives at recommendations, they can integrate insights with judgment, creativity, and moral reasoning. Misalignment or opacity undermines trust and reduces the benefits of augmentation. Intentional design ensures that AI acts as an enabler of human intelligence rather than a source of confusion or dependence. Flourishing emerges when humans guide AI purposefully, maintaining agency and insight.

Augmented intelligence also reshapes education and skill development. Humans must learn to interpret AI outputs, integrate insights with ethical judgment, and communicate effectively with machine-assisted systems. Skills such as critical thinking, creativity, problem-solving, and emotional intelligence become central. AI handles data processing and pattern recognition, freeing humans to focus on tasks requiring interpretation, collaboration, and value-based judgment. Flourishing is realized when humans cultivate capacities that AI cannot replicate, using technology to enhance depth and effectiveness.

Augmented intelligence also enhances creativity across disciplines. AI can generate ideas, identify novel connections, and explore vast possibilities, expanding the human imagination. Yet meaning, context, and intentionality remain human responsibilities. Artists, writers, designers, and innovators use AI as a tool to extend creative reach, not to replace their unique insight. Collaboration between human intuition and machine computation generates outcomes richer than either could produce alone. Flourishing arises when humans harness AI to deepen creativity while retaining authorship and ethical direction.

In decision-making, augmented intelligence strengthens judgment under complexity. AI can analyze massive datasets, simulate scenarios, and identify risks or opportunities that might elude humans. However, humans remain responsible for evaluating outcomes, weighing trade-offs, and making choices aligned with ethical principles and long-term objectives. Augmentation ensures that decisions are informed, not automated, combining machine precision with human insight. Flourishing is realized when humans maintain authority over ethical, strategic, and relational dimensions while benefiting from computational power.

Augmented intelligence reshapes professional identity and skill requirements. As AI assumes routine, analytical, or computational tasks, humans are freed to focus on interpretation, leadership, and ethical judgment. Expertise evolves to include collaboration with intelligent systems, combining technical literacy with cognitive flexibility and moral discernment. Workers who adapt to this paradigm leverage AI to enhance productivity, creativity, and problem-solving. Flourishing arises when humans recognize their irreplaceable contributions in areas where intelligence, judgment, and relational understanding are required.

Healthcare provides a vivid example of augmented intelligence in practice. AI can assist in diagnosing complex conditions, predicting outcomes, and personalizing treatment plans. Yet human physicians interpret findings, communicate with patients, and make ethically grounded decisions. AI acts as a powerful extension of human capability, enabling better care without replacing the human touch. Flourishing in such contexts depends on integrating technological insights with empathy, ethical responsibility, and relational competence.

Augmented intelligence also transforms learning and education. AI tutors, adaptive learning systems, and personalized analytics can provide students with real-time feedback, tailored challenges, and data-driven insights. However, understanding, critical thinking, ethical judgment, and the cultivation of wisdom remain human responsibilities. Teachers guide interpretation, moral reasoning, and social-emotional development. Flourishing arises when students learn to collaborate with AI, using technology to extend their cognitive and creative abilities while maintaining human judgment and reflection.

The workplace increasingly reflects the principles of augmented intelligence. Professionals in finance, engineering, research, and creative industries rely on AI to analyze complex datasets, forecast trends, and generate possibilities. Humans then apply contextual knowledge, ethical reasoning, and strategic judgment to determine next steps. Collaboration ensures that AI serves as an enabler rather than a replacement, magnifying human impact without diminishing responsibility. Flourishing depends on cultivating capacities that AI cannot replicate, emphasizing insight, ethics, and creativity over mere output.

Ethical oversight is central to augmented intelligence. Humans determine the values, goals, and boundaries guiding AI deployment. Machines may offer recommendations or predictions, but they lack moral awareness, empathy, and accountability. Flourishing arises when humans take responsibility for the ethical integration of AI, ensuring that technology aligns with societal well-being, fairness, and justice. Augmentation is effective only when paired with human moral stewardship, preventing harm while maximizing benefit.

Collaboration with AI also changes problem-solving methodologies. Complex challenges benefit from human intuition and creativity paired with machine computation, modeling, and pattern recognition. AI can simulate countless scenarios, but humans provide judgment, context, and prioritization. Together, humans and machines can tackle tasks that neither could manage alone, from climate modeling to medical research. Flourishing emerges from strategic collaboration, where human intelligence guides and interprets AI-generated possibilities.

The promise of augmented intelligence lies not in efficiency alone, but in human empowerment. AI provides insights, generates possibilities, and streamlines processes, but humans decide how to act, what values to prioritize, and which paths to pursue. Flourishing arises when humans retain agency, moral oversight, and creative authorship, using AI to extend their capabilities rather than outsource judgment. This partnership transforms not just work, but identity, purpose, and societal impact.

Augmentation requires continuous learning and adaptation. Humans must cultivate skills that complement AI—critical thinking, ethical reasoning, creativity, and emotional intelligence. Education and professional development must emphasize collaboration with intelligent systems, not competition with them. By understanding AI's strengths and limits, humans can maximize the potential of co-creation. Flourishing emerges from this ongoing interplay between human insight and machine capability.

Ethical responsibility is central to the augmented intelligence paradigm. Humans define objectives, boundaries, and acceptable risks. Machines may suggest options, simulate scenarios, or highlight probabilities, but ethical judgment and accountability remain human domains. Flourishing depends on deliberate stewardship, ensuring that AI enhances collective well-being, fairness, and justice. Augmentation is not just functional—it is moral, requiring thoughtful guidance and principled action.

Augmented intelligence also transforms creativity, innovation, and problem-solving. AI can rapidly generate possibilities, identify unseen patterns, and model complex systems, expanding the horizon of human imagination. Humans provide context, evaluation, and ethical discernment, transforming raw computational outputs into meaningful solutions. Flourishing arises when collaboration is intentional, principled, and guided by human values, producing outcomes that are both innovative and ethically grounded.

Finally, augmented intelligence reaffirms what it means to be human. Machines amplify capability, but they cannot replicate judgment, empathy, or moral insight. Humans remain the ultimate authors of purpose, ethics, and creativity. Flourishing occurs when humans embrace AI as a partner, leveraging its strengths while guiding it with wisdom, empathy, and ethical responsibility.

Augmentation transforms fear into opportunity, positioning humans to thrive in a world of expanding technological potential.

Author Reflection:

Writing this chapter reinforced the idea that AI's greatest value lies in partnership, not replacement. Machines enhance human capabilities, but judgment, ethics, and creativity remain distinctly human. I realized that flourishing depends on cultivating skills, insight, and moral responsibility while leveraging AI as an amplifier of potential. Augmented intelligence is not a threat—it is an invitation to expand what humans can achieve ethically, creatively, and collectively.

Inspirational Quote:

"*AI is not a replacement for human wisdom, but a partner that magnifies our creativity, judgment, and ethical responsibility.*" — Anonymous

Chapter . 12

Decision-Making in the Age of Algorithms

Human judgment remains essential even as machines optimize choices.

In the modern world, algorithms increasingly shape the decisions humans make. From finance and healthcare to hiring and law enforcement, AI-driven systems influence outcomes with speed, precision, and scale. Yet the use of algorithms does not eliminate the need for human judgment; rather, it redefines its role. Humans must interpret, contextualize, and ethically evaluate algorithmic outputs to ensure decisions align with societal and moral values. Flourishing arises from combining computational power with discernment, intuition, and ethical awareness. Technology can enhance choices, but it cannot replace responsibility.

Algorithms excel at processing data, identifying patterns, and predicting outcomes. They can optimize efficiency, reduce error, and highlight probabilities that would elude human perception. Yet algorithms operate within the parameters set by humans, and their outputs reflect the biases and assumptions embedded in their design. Human oversight is essential to detect unintended consequences, evaluate fairness, and prioritize ethical considerations. Flourishing depends on humans guiding algorithmic power with critical thinking, moral discernment, and social awareness.

Decision-making is fundamentally a human act, informed but not dictated by AI. Algorithms provide options, forecasts, or recommendations, but humans determine which course to pursue based on values, context, and consequences. In high-stakes domains such as healthcare, criminal justice, and environmental policy, ethical and moral judgment is indispensable. Machines can assist, but accountability and responsibility remain human. Flourishing arises when humans integrate insight, ethics, and relational awareness with algorithmic analysis to make informed, responsible decisions.

The rise of algorithmic influence also introduces new forms of uncertainty and bias. AI systems can perpetuate inequities or reinforce systemic patterns if left unchecked. Humans must critically evaluate the assumptions, data quality, and potential biases inherent in AI recommendations. Decision-making becomes an exercise in ethical vigilance, balancing efficiency with fairness, and probability with human values. Flourishing requires awareness of these dynamics and deliberate action to ensure equitable outcomes.

The integration of algorithms into decision-making requires transparency and interpretability. Humans must understand how AI systems arrive at recommendations, including the data inputs, assumptions, and logic behind outputs. Black-box systems may provide efficient answers, but without comprehension, humans risk blindly following potentially flawed or biased guidance. Flourishing arises when humans combine algorithmic insight with comprehension, critique, and ethical judgment. Transparency enables accountability, trust, and responsible action in complex decision-making environments.

Algorithmic decision-making also changes the pace of human choices. In areas like financial trading, emergency response, and logistics, algorithms operate at speeds far beyond human capacity. While this acceleration increases efficiency, it also challenges humans to maintain oversight and ethical reflection. Decisions must balance speed with deliberation, considering long-term consequences, fairness, and social impact. Flourishing depends on humans exercising discernment, even when machines enable rapid computation, ensuring that ethical considerations are not sacrificed for efficiency.

Collaboration between humans and algorithms can enhance creativity in problem-solving. Algorithms can identify patterns, simulate scenarios, and generate options that expand the range of potential solutions. Humans interpret, evaluate, and select approaches aligned with values, context, and strategic objectives. Flourishing arises when this partnership produces solutions that are both innovative and ethically grounded. Human judgment adds purpose, meaning, and moral clarity to algorithmic efficiency, creating outcomes neither could achieve alone.

Critical thinking becomes more essential than ever. The availability of algorithmic recommendations can create a false sense of certainty or overreliance. Humans must question assumptions, evaluate alternatives, and weigh ethical considerations before acting. Algorithmic outputs are tools, not mandates; judgment remains a human responsibility. Flourishing emerges when individuals develop discernment, ethical awareness, and the ability to synthesize human insight with computational analysis. Intelligent decision-making balances efficiency with responsibility, insight with morality.

Algorithmic influence also impacts personal decision-making. From career choices to financial planning, AI-powered tools provide recommendations, forecasts, and risk assessments. While these tools enhance information access, humans remain responsible for interpreting outputs, weighing priorities, and making value-driven choices. Blind reliance on algorithms can erode agency and critical thinking, whereas thoughtful integration amplifies insight and effectiveness. Flourishing occurs when humans use AI to support informed, ethical, and reflective decision-making, rather than outsourcing responsibility entirely.

Organizations must cultivate a culture that balances human judgment with algorithmic efficiency. Employees should be empowered to question, interpret, and adapt machine-generated insights to organizational goals and ethical standards. Leadership plays a critical role in promoting ethical engagement, oversight, and reflection in algorithm-driven processes. Flourishing organizations recognize that human responsibility cannot be delegated to machines, and decision-making benefits from both human discernment and computational support.

Algorithmic decision-making also affects public policy and governance. Governments increasingly use AI for predictive analytics, resource allocation, and social services optimization. While algorithms can provide efficiency and scale, they cannot ensure fairness, justice, or equity. Humans must interpret results, set policy priorities, and safeguard against bias or unintended consequences. Flourishing societies balance computational power with ethical governance, using AI as a tool for informed and responsible policymaking.

The risks of algorithmic overreach highlight the importance of human oversight. AI systems can perpetuate discrimination, reinforce stereotypes, or misinterpret context if humans do not monitor and guide outcomes. Ethical review, accountability structures, and continuous evaluation are essential to prevent harm. Flourishing depends on maintaining vigilance, exercising moral judgment, and ensuring AI supports values of equity, transparency, and human dignity. Decision-making remains a deeply human responsibility, augmented—not replaced—by technology.

Algorithmic decision-making also reshapes risk management. Machines can analyze probabilities, model scenarios, and forecast outcomes more comprehensively than humans alone. Yet humans interpret these predictions, weigh ethical implications, and decide which risks are acceptable. Overreliance on algorithms may lead to moral or strategic blind spots, while thoughtful integration enhances judgment, foresight, and societal benefit. Flourishing arises when humans combine computational power with ethical reasoning, prudence, and long-term vision in managing uncertainty.

Collaboration between humans and algorithms enables more informed strategic planning. Leaders can explore multiple scenarios, evaluate potential outcomes, and consider consequences across ethical, financial, and social dimensions. Machines provide clarity and data-driven insights, but humans provide context, values, and direction. Flourishing occurs when humans actively guide AI outputs, ensuring that decisions align with broader human objectives, social justice, and sustainable outcomes.

Algorithmic decision-making also transforms organizational efficiency. AI can optimize workflows, allocate resources, and identify operational improvements. However, human oversight ensures that optimization aligns with ethical standards and long-term objectives. Efficiency gains are meaningful only when paired with human judgment, relational intelligence, and ethical responsibility. Flourishing emerges from balancing machine-optimized performance with human-centered values and moral considerations.

Decision-making in the age of algorithms also requires adaptability. Algorithms evolve, data changes, and societal expectations shift, demanding continuous human assessment and recalibration. Humans must maintain awareness of system updates, limitations, and potential biases to ensure responsible outcomes. Flourishing depends on vigilance, flexibility, and integration of human judgment with technological assistance. Collaboration with AI enhances problem-solving while preserving accountability and ethical oversight.

Decision-making in the age of algorithms ultimately highlights the enduring importance of human values. AI systems optimize, predict, and analyze, but they cannot determine what is fair, just, or meaningful. Humans remain responsible for interpreting outputs, prioritizing goals, and ensuring decisions align with ethical principles. Flourishing depends on recognizing that technology is a tool for informed action, not a substitute for judgment. Human oversight ensures that algorithmic power supports collective well-being, equity, and purpose.

Algorithmic systems also amplify both the potential and the consequences of human decisions. Errors or biases in design can propagate widely and rapidly, impacting individuals and communities. Human judgment, ethical reasoning, and continuous monitoring are essential safeguards. Flourishing arises when humans actively guide AI, critically evaluating outcomes and making value-driven choices. Responsibility, discernment, and moral clarity remain irreplaceable in an increasingly algorithm-driven world.

Collaboration with algorithms can enhance creativity and innovation in decision-making. Machines can generate possibilities, simulate scenarios, and analyze trends, expanding the decision space. Humans provide interpretation, context, and ethical guidance, shaping outputs into actionable, responsible, and meaningful decisions. Flourishing occurs when human judgment and technological insight are integrated intentionally, producing outcomes that are both effective and aligned with societal values.

Education and awareness are essential for human flourishing in an algorithmic era. Understanding how algorithms function, recognizing limitations, and cultivating ethical reasoning equip humans to guide technology effectively. Critical thinking and moral discernment become indispensable skills, ensuring that decisions made with AI support are fair, responsible, and beneficial. Flourishing arises when humans leverage algorithms to enhance insight without surrendering accountability or ethical authority.

Finally, decision-making in the age of algorithms is an invitation to exercise human wisdom at scale. AI magnifies our reach, efficiency, and analytical capacity, but purpose, judgment, and ethical oversight remain human responsibilities. Flourishing occurs when humans embrace their role as stewards of both technology and consequence, guiding AI to serve insight, justice, and collective well-being. The age of algorithms is not a threat to human judgment—it is an opportunity to exercise it more effectively and ethically.

Author Reflection:

Writing this chapter emphasized that algorithms are tools, not arbiters of morality or meaning. Humans remain responsible for interpretation, judgment, and ethical oversight in every decision. I realized that flourishing depends on cultivating critical thinking, moral discernment, and adaptability alongside technological literacy. Algorithms can expand insight and efficiency, but wisdom, fairness, and accountability remain uniquely human. Decision-making is at its best when humans guide machines with purpose, ethics, and reflection.

Inspirational Quote:

"Algorithms can inform, optimize, and predict—but human judgment, guided by ethics and wisdom, shapes meaningful action." — Anonymous

Chapter . 13

Trust, Transparency, and Control

Healthy collaboration requires understanding how machines think and act.

Trust is the foundation of any successful collaboration between humans and AI. Without trust, humans may reject machine recommendations or fail to use technology effectively. Conversely, overreliance on AI can lead to blind acceptance of outputs without critical evaluation. Trust emerges when humans understand how AI functions, its limitations, and the assumptions embedded in its design. Flourishing depends on balancing confidence with discernment, ensuring that collaboration enhances rather than diminishes human judgment. Ethical oversight, clear communication, and accountability are essential for cultivating trust in AI systems.

Transparency is central to building trust. Humans must be able to see, understand, and evaluate how AI systems arrive at decisions. Explainable algorithms, open data practices, and interpretable models empower users to make informed judgments about the reliability and relevance of outputs. Opaque "black-box" systems can undermine confidence, perpetuate errors, and amplify bias. Flourishing arises when humans can scrutinize AI processes, integrating machine insight with ethical reflection and contextual understanding. Transparency transforms AI from a mysterious tool into a reliable partner.

Control is another crucial dimension of collaboration. Humans must retain authority over objectives, decision thresholds, and ethical boundaries. AI can optimize and recommend, but it cannot assume moral responsibility or prioritize long-term social values. Control ensures that machines serve human goals rather than dictate them. Flourishing occurs when humans actively guide AI, setting parameters, monitoring outcomes, and intervening when necessary to uphold ethical and practical standards. Partnership is meaningful only when human oversight remains central.

Trust, transparency, and control are interconnected. Transparent systems increase understanding, which fosters trust, and trust allows humans to exercise control responsibly. Neglecting any of these elements can undermine collaboration, resulting in misuse, ethical lapses, or loss of confidence. Flourishing arises when humans cultivate all three elements, creating a collaborative ecosystem in which AI amplifies human capacity without compromising judgment or values. Healthy systems prioritize human agency, clarity, and accountability.

Trust also depends on consistent and reliable performance. AI systems must function predictably and provide outputs that align with user expectations and contextual reality. Unexpected or inconsistent behavior erodes confidence and hinders collaboration. Humans assess reliability not only through technical accuracy but also through ethical alignment and social responsibility. Flourishing arises when AI performance is transparent, consistent, and accountable, reinforcing trust while enhancing human decision-making.

Transparency extends to the underlying data used by AI systems. Humans must understand how data is collected, processed, and interpreted, as data quality directly affects outcomes. Bias, incompleteness, or misrepresentation in data can distort decisions, even when algorithms function correctly. Open access to data provenance, validation processes, and potential limitations allows humans to evaluate AI outputs critically. Flourishing depends on transparency in both the algorithm and the data feeding it, ensuring informed judgment and ethical oversight.

Control involves the ability to intervene, override, or adjust AI recommendations when necessary. Humans must maintain agency in critical decisions, particularly in high-stakes domains such as healthcare, criminal justice, or financial regulation. Control is not about micromanaging every algorithmic step, but about ensuring alignment with human values, ethical principles, and long-term goals. Flourishing emerges when humans exercise authority responsibly, using AI as an amplifier rather than a substitute for judgment or moral accountability.

Trust, transparency, and control are also relational and organizational challenges. Teams, institutions, and stakeholders must collaborate to define standards, monitor compliance, and ensure accountability. Transparent governance structures and clear responsibilities reinforce confidence in AI systems. Humans must coordinate ethical oversight, technical understanding, and social responsibility to maintain balance. Flourishing depends on collaborative stewardship, where humans collectively uphold the integrity of AI-guided decision-making.

Building trust also involves managing expectations. Humans often overestimate AI's capabilities, assuming intelligence, creativity, or moral judgment where none exist. Conversely, fear of error may lead to underutilization of valuable insights. Transparent communication about AI's strengths, limitations, and intended purposes calibrates expectations and fosters confidence. Flourishing emerges when humans understand what AI can and cannot do, leveraging its capabilities while maintaining agency and ethical responsibility.

Transparency in design and operation strengthens collaboration. Users should have access to explanations of algorithmic logic, decision pathways, and key assumptions. Explainable AI empowers humans to interrogate results, challenge outputs, and ensure alignment with values. Opaque systems undermine trust, potentially leading to misuse or overreliance. Flourishing depends on visibility into both process and reasoning, enabling informed decisions and ethical oversight.

Control also involves establishing boundaries for AI intervention. Humans must define where machine recommendations end and human decision-making begins. Clear guidelines, ethical guardrails, and decision thresholds ensure that AI acts as an assistant rather than an autonomous authority. Flourishing arises when humans maintain ultimate responsibility, integrating AI outputs with ethical judgment, context, and societal priorities.

Organizational culture plays a critical role in trust, transparency, and control. Teams that prioritize ethical AI deployment, encourage questioning, and foster collaboration create an environment where technology enhances rather than undermines human decision-making. Leadership must model ethical stewardship, accountability, and transparency to cultivate confidence and collaboration. Flourishing emerges when human oversight, social norms, and technological design align to support ethical, effective, and responsible action.

Trust in AI is deeply tied to accountability. Humans must be able to identify who is responsible when AI-driven decisions lead to errors or unintended consequences. Clear lines of accountability ensure that ethical standards are upheld and that AI remains a tool under human stewardship rather than an autonomous actor. Flourishing arises when humans embrace responsibility, maintaining oversight and ethical governance over AI systems while leveraging their capabilities.

Transparency also requires openness about potential biases and limitations. Algorithms reflect the assumptions, design choices, and data inputs of their creators. Biases in training data or model architecture can produce skewed or unfair outcomes if unaddressed. Human oversight is essential to identify, mitigate, and correct these biases. Flourishing occurs when humans actively interrogate AI systems, ensuring decisions support fairness, justice, and social responsibility rather than perpetuating inequities.

Control extends beyond immediate decision-making to long-term strategy. Humans must define objectives, set ethical priorities, and determine how AI aligns with broader societal goals. Without intentional guidance, AI can optimize for narrow metrics—such as efficiency or profitability—while neglecting human values and well-being. Flourishing is achieved when humans retain authority over the purposes, boundaries, and ethical constraints guiding AI deployment.

Collaboration with AI also requires cultivating understanding across diverse stakeholders. Users, developers, and policymakers must communicate effectively to ensure that systems are trustworthy, transparent, and controllable. Shared knowledge and clear governance structures build collective confidence, reduce misuse, and enhance ethical integration. Flourishing arises from a culture of shared responsibility, where humans collectively steward AI to amplify human potential while safeguarding societal values.

Concluding the discussion on trust, transparency, and control, it is clear that human agency remains paramount. Machines can provide recommendations, analyze data, and highlight possibilities, but responsibility, ethical judgment, and accountability always rest with humans. Flourishing is realized when humans actively engage with AI, setting boundaries, scrutinizing outputs, and maintaining oversight to ensure alignment with values and societal good. Technology is a partner, not a replacement, in decision-making and ethical action.

Trust develops gradually through experience and understanding. Human users must interact with AI repeatedly, observing reliability, accuracy, and ethical alignment. Transparency and control reinforce trust by clarifying processes, assumptions, and limitations. When humans know what to expect from AI and how to intervene if necessary, collaboration becomes more effective and ethically sound. Flourishing emerges from a partnership grounded in comprehension, confidence, and human-centered design.

Ethical reflection remains central. Humans must question not only AI outputs but also the objectives and priorities embedded in their design. Who benefits from AI recommendations? What values are being promoted or neglected? What risks or inequities might be amplified? Flourishing occurs when humans answer these questions deliberately, guiding AI toward outcomes that enhance well-being, equity, and collective good.

Collaboration with AI also highlights the importance of education and literacy. Users must understand how algorithms function, recognize biases, and cultivate the ability to intervene responsibly. Transparency, trust, and control are not innate—they require ongoing learning, vigilance, and ethical engagement. Flourishing depends on humans developing both technical competence and moral discernment to steward AI effectively.

Finally, the triad of trust, transparency, and control embodies the principle that humans remain at the center of AI integration. Machines amplify capabilities, but humans provide judgment, ethical oversight, and purpose. Flourishing arises when AI is guided consciously, responsibly, and collaboratively, ensuring that technology serves insight, equity, and human well-being rather than replacing the uniquely human qualities that sustain a just and flourishing society.

Author Reflection:

Writing this chapter reinforced my belief that trust, transparency, and control are inseparable in human–AI collaboration. Humans remain the moral, ethical, and judgmental anchors guiding technology. I realized that flourishing depends not only on technical understanding but also on vigilance, ethical reflection, and proactive stewardship. Machines amplify human capabilities, but humans determine their purpose, ensuring that AI serves societal well-being and collective growth.

Inspirational Quote:

"*Machines can perform tasks, but only humans can guide them with trust, transparency, and ethical purpose*." — Anonymous

Part IV — Ethics, Power, and Inequality

Chapter . 14

Who Owns the Machines That Shape Reality

Control over AI systems determines who benefits and who is left behind.

Ownership of AI is no longer just a legal or economic question—it is a question of power. Those who control AI systems influence decisions, shape opportunities, and affect the distribution of wealth and knowledge. Machines capable of analyzing, predicting, and generating insights can significantly alter markets, governance, and social outcomes. Whoever controls these systems holds leverage over societal realities, including access to information, employment, and influence. Flourishing depends on understanding the implications of ownership, governance, and equitable access. Technology is not neutral; its impact is shaped by those who design, deploy, and control it.

Control over AI also dictates who can innovate and who is constrained. Corporations, governments, and research institutions with resources to develop or acquire AI gain disproportionate advantage. Smaller organizations, individuals, and marginalized communities may lack access, widening social and economic disparities. Human flourishing requires consideration of access, inclusivity, and democratization of AI. Power imbalances are not inherently technological—they are human decisions reflected in ownership and distribution. Understanding this dynamic is crucial for creating ethical and equitable systems.

Ownership impacts transparency and accountability. Those who design and control AI determine what information is accessible, how decisions are explained, and who can intervene. Without shared oversight, algorithms can reinforce inequities, obscure responsibility, and concentrate power. Humans must consider ownership not just in financial terms, but in ethical, social, and governance dimensions. Flourishing emerges when communities, policymakers, and developers collaborate to ensure that AI serves collective well-being rather than narrow interests.

Economic implications are profound. AI-driven automation and augmentation shift the distribution of wealth and labor. Owners of AI systems may capture the majority of economic benefits, while workers and communities face disruption or exclusion. Decisions about ownership and access shape who prospers and who is left behind. Flourishing requires mechanisms to distribute AI benefits widely, promoting opportunity, dignity, and shared prosperity. Control over technology determines the trajectory of human and societal outcomes.

Ownership of AI extends beyond corporations to governments and public institutions. Governments wield AI in areas like law enforcement, social services, and national security, shaping citizen experiences and societal norms. Without careful regulation and oversight, centralized control can lead to abuse, inequity, and loss of public trust. Flourishing depends on establishing transparent, accountable frameworks that balance innovation, efficiency, and democratic values. Ownership is not merely possession—it is responsibility for the societal consequences of AI deployment.

Access and inclusion are central ethical considerations in ownership. Communities that lack access to AI tools are disadvantaged in education, employment, and civic participation. Marginalized populations, lower-income groups, and smaller organizations often face systemic barriers to AI adoption. Ensuring equitable access requires deliberate policies, investment in public infrastructure, and inclusive design practices. Flourishing arises when AI ownership is paired with initiatives that democratize access and empower a broad spectrum of society to benefit from technological advancement.

Intellectual property and licensing also shape the landscape of ownership. Patents, proprietary systems, and closed-source platforms can consolidate power and limit collaboration. Open-source and community-driven AI projects offer alternative models that foster innovation, transparency, and shared benefit. Humans must weigh the trade-offs between commercial incentives and collective welfare. Flourishing emerges when ownership models align with societal values, promoting innovation while ensuring ethical, equitable outcomes.

Ownership also influences accountability and oversight. Those who control AI systems determine how errors, biases, and harms are addressed. Without shared governance, responsibility can be obscured, leaving affected individuals and communities vulnerable. Humans must design mechanisms to distribute accountability fairly, ensuring that owners cannot evade ethical obligations. Flourishing depends on ownership models that combine control with transparency, oversight, and social responsibility.

Ownership shapes the economic dynamics of innovation. Entities that control AI can capture the majority of financial returns, leaving smaller competitors and the broader workforce at a disadvantage. This concentration of power can limit diversity in research, creativity, and entrepreneurial activity. Human flourishing depends on ownership structures that enable broader participation, fair compensation, and shared opportunity. Technology becomes a tool for empowerment rather than a mechanism for inequality when ownership is balanced with inclusivity and social responsibility.

Control over AI systems also affects societal decision-making. Algorithms influence access to credit, healthcare, education, and employment opportunities. Owners of AI systems determine priorities, weighting, and outcomes, effectively shaping life chances for millions. Ethical ownership requires attention to fairness, accountability, and equity, ensuring that AI contributes to societal good rather than perpetuating systemic inequities. Flourishing emerges when those in control consciously act to mitigate harm and expand opportunity.

Transparency is deeply connected to ownership. When AI is owned and managed by a select few, transparency may be limited, preventing oversight and public understanding. Open access to AI processes, datasets, and design decisions fosters trust and encourages responsible use. Humans benefit most when ownership is accompanied by policies that prioritize clarity, accountability, and ethical governance. Flourishing depends on shared knowledge and collaborative stewardship, allowing society to guide the development and application of AI.

Ownership also affects cultural and social narratives. AI systems can shape information consumption, media representation, and public discourse. Those in control determine which voices are amplified and which are marginalized. Human flourishing requires that ownership structures consider these sociocultural impacts, promoting inclusivity and resisting monopolization of influence. Ethical stewardship ensures that AI contributes to a diverse, informed, and equitable public sphere.

Ownership and control also shape the pace and direction of innovation. Entities with substantial resources can dictate research priorities, focusing on profit-driven or politically advantageous applications. Meanwhile, areas with less commercial appeal—such as public health, environmental sustainability, or education—may receive insufficient attention. Humans must advocate for ownership models that align technological development with societal needs. Flourishing depends on balancing innovation incentives with public benefit, ensuring that AI serves broad human interests.

Equity in ownership is critical for addressing systemic disparities. Historically marginalized communities risk being excluded from the benefits of AI due to limited access, representation, or decision-making power. Deliberate efforts to democratize ownership—through public investment, cooperative models, or inclusive governance—can counteract these inequities. Flourishing arises when AI ownership is structured to provide opportunity, empowerment, and participation for all segments of society.

Global considerations complicate ownership further. Multinational corporations and cross-border collaborations create networks of influence that can concentrate control in specific regions or institutions. Nations and populations with less technological capacity may become dependent on AI systems developed elsewhere, reinforcing global inequalities. Ethical ownership requires international cooperation, knowledge-sharing, and equitable access. Flourishing emerges when ownership frameworks consider both local and global consequences, ensuring fairness, inclusion, and shared prosperity.

Transparency in decision-making and accountability mechanisms are inseparable from ownership. Owners must be answerable for the consequences of AI deployment, from societal impacts to ethical breaches. Legal, regulatory, and social frameworks help ensure that those in control act responsibly, mitigating harm and amplifying collective benefit. Flourishing depends on a governance ecosystem where ownership includes both privilege and responsibility, aligning AI use with human values and long-term societal goals.

Concluding the discussion on ownership, it is clear that control over AI determines who benefits and who is left behind. Machines do not act independently; humans define their purpose, priorities, and reach. Ownership is not simply legal possession—it is a moral and societal responsibility. Flourishing arises when ownership structures ensure that AI is accessible, accountable, and aligned with collective human well-being rather than narrow commercial or political interests.

Ownership also influences the ethical deployment of AI. Those who control systems dictate standards for fairness, privacy, and transparency. Concentrated ownership without oversight risks amplifying bias, reinforcing inequality, and limiting opportunity. Conversely, equitable and responsible ownership can democratize access, foster innovation, and protect vulnerable populations. Flourishing depends on owners exercising stewardship that balances innovation with ethics, efficiency with social responsibility.

Transparency and governance are essential complements to ownership. Clear documentation of decision-making processes, open communication about capabilities and limitations, and mechanisms for public accountability enhance trust. Flourishing occurs when ownership is paired with transparency, ethical oversight, and participatory frameworks that empower stakeholders to guide AI's impact. Control without accountability undermines societal confidence and ethical integrity.

Global and societal equity must be central to ownership considerations. Unequal distribution of AI systems across nations, communities, and demographics risks creating new hierarchies of power. Deliberate policies, cooperative frameworks, and inclusive design practices can redistribute access and influence. Flourishing emerges when ownership models prioritize fairness, inclusion, and the shared benefits of AI advancement.

Finally, ownership is ultimately about responsibility and stewardship. Humans, not machines, shape societal outcomes, ethical norms, and opportunities through their control of AI. Flourishing occurs when ownership is exercised thoughtfully, inclusively, and ethically, ensuring that technology empowers rather than marginalizes. Control is privilege, but it is also an obligation to guide AI in service of humanity's collective well-being.

Author Reflection:

Writing this chapter underscored that ownership of AI is inseparable from responsibility. Control shapes access, equity, and societal impact. I realized that flourishing depends not just on who owns technology, but on how that ownership is exercised—ethically, transparently, and inclusively. AI can magnify human potential or inequality, and humans must consciously choose the path that empowers collective well-being.

Inspirational Quote:

"*Ownership of technology is power, but the true measure lies in how that power is wielded for collective human flourishing.*" — Anonymous

Chapter . 15

Bias in Code and the Reinvention of Discrimination

AI can reinforce inequality unless consciously designed to resist it.

Bias in AI is not an abstract problem—it is a reflection of human decisions embedded in code. Algorithms are trained on data generated by human society, which is often biased, incomplete, or unrepresentative. These biases can be amplified, reproduced, and institutionalized when algorithms make decisions affecting hiring, lending, policing, or access to resources. Humans must confront the reality that technology is not neutral; its outputs reflect societal structures, values, and assumptions. Flourishing arises when AI is consciously designed, monitored, and corrected to resist discrimination and promote equity. Ignoring bias risks perpetuating historical inequities in new, technologically mediated forms.

The reinvention of discrimination through code highlights how subtle bias can be embedded in decision-making processes. Algorithmic outputs may appear objective, but design choices—including which variables to consider, how to weight data, and which historical outcomes to prioritize—can reproduce systemic inequality. Bias can operate unintentionally, yet its consequences are real and far-reaching. Flourishing depends on human vigilance, ethical design, and active mitigation strategies that prevent AI from reinforcing harmful patterns. Technology must be guided by deliberate reflection, not assumed neutrality.

Understanding bias requires examining both data and context. Historical data may reflect systemic discrimination, while underrepresentation can skew results for marginalized communities. AI models trained on such data can unintentionally disadvantage certain groups, even when algorithms are technically correct. Humans must interpret results critically, evaluate context, and consider fairness beyond mere accuracy. Flourishing emerges when humans use insight, ethics, and social awareness to ensure AI outcomes enhance equity rather than replicate bias.

Bias in AI is not limited to individual decisions—it can affect entire systems. Employment algorithms, credit scoring, and predictive policing tools can influence access to opportunity, wealth, and social mobility. Minor coding decisions can have cascading effects across populations, creating new forms of structural inequality. Humans must engage in proactive oversight, transparent design, and inclusive evaluation to ensure systems do not entrench disadvantage. Flourishing arises when humans recognize and correct systemic bias, leveraging AI to advance fairness and social good.

Bias in AI also raises questions about accountability. When discriminatory outcomes occur, it is essential to determine who is responsible—the developers, the organizations deploying the system, or the societal structures reflected in the data. Clear accountability mechanisms ensure that harm can be addressed, corrected, and prevented in the future. Flourishing depends on establishing responsibility frameworks that hold humans—not machines—accountable for ethical outcomes. AI magnifies impact, but moral and legal oversight must remain human-centered.

Inclusivity in design is a powerful tool against bias. Involving diverse teams in AI development—across gender, race, socioeconomic status, and cultural background—reduces blind spots and ensures multiple perspectives shape algorithms. Inclusion helps identify potential ethical concerns and contextual nuances that may be overlooked by homogeneous teams. Flourishing emerges when diversity informs both the creation and evaluation of AI, producing systems that respect equity and human dignity.

Transparency is equally critical. Humans must understand how algorithms operate, what data informs them, and which assumptions influence their decisions. Explainable AI allows users and stakeholders to scrutinize outputs, identify biases, and intervene when necessary. Without transparency, discrimination can remain hidden, reinforcing inequality in subtle but significant ways. Flourishing arises when humans can interrogate, guide, and correct AI, ensuring outputs align with fairness and justice.

Education and awareness are central to addressing bias. Developers, users, policymakers, and the public must understand the potential for AI to reproduce discrimination. Ethical literacy, data literacy, and critical thinking equip humans to evaluate algorithmic outcomes responsibly. Flourishing occurs when society cultivates informed oversight, ensuring that technology serves equitable and just ends rather than amplifying harm.

Bias in AI has tangible consequences across daily life. Hiring algorithms may favor certain demographics, while lending systems can systematically deny loans to marginalized communities. Predictive policing tools may disproportionately target neighborhoods with historical over-policing, reinforcing cycles of disadvantage. Even seemingly neutral decisions, when shaped by biased data or flawed assumptions, can perpetuate structural inequities. Flourishing arises when humans actively recognize these patterns, question automated outcomes, and intervene to prevent harm.

Mitigating bias requires intentional technical strategies. Techniques such as data balancing, fairness constraints, and adversarial testing can reduce discriminatory outputs. However, these methods alone are insufficient without human ethical oversight. Technical solutions must be coupled with reflection on societal values, context, and consequences. Flourishing emerges when humans combine computational methods with ethical reasoning, ensuring that AI systems enhance opportunity and fairness rather than undermine them.

Bias in AI also intersects with power and influence. Those who control algorithmic systems can shape outcomes that affect social mobility, education, and civic participation. Marginalized groups may experience exclusion, misrepresentation, or systemic disadvantage if algorithms are left unchecked. Ethical ownership, inclusive design, and participatory governance are essential for counteracting these inequities. Flourishing arises when AI is developed and deployed to empower communities, reduce disparity, and promote social justice.

Legal and regulatory frameworks play a critical role in addressing algorithmic bias. Policies that enforce transparency, auditing, and accountability ensure that organizations are responsible for discriminatory outcomes. Human oversight remains essential; laws provide structure, but ethical commitment guides practice. Flourishing depends on creating ecosystems where AI is governed by both rules and moral principles, ensuring technology aligns with societal values and human rights.

Addressing bias requires a culture of ongoing reflection and accountability. AI systems are not static; they evolve as data changes, user behavior shifts, and new applications emerge. Continuous monitoring ensures that discriminatory patterns do not reappear or intensify over time. Humans must evaluate outcomes regularly, update models responsibly, and intervene when biases arise. Flourishing depends on proactive vigilance, combining technical expertise with ethical judgment to safeguard fairness.

Collaboration across disciplines strengthens efforts to mitigate bias. Engineers, ethicists, sociologists, policymakers, and community representatives bring diverse perspectives that illuminate potential pitfalls. By integrating technical, ethical, and social insight, AI systems can be designed to reflect broader human values. Flourishing emerges when cross-disciplinary collaboration ensures that algorithms respect fairness, inclusivity, and social responsibility.

Education and empowerment of users are also critical. People interacting with AI must understand its limitations, potential biases, and areas of uncertainty. Awareness enables critical evaluation, appropriate skepticism, and ethical intervention. Flourishing occurs when humans are equipped to question, guide, and influence AI outputs responsibly, rather than accepting automated decisions uncritically.

Bias in AI underscores the importance of participatory governance. Communities affected by algorithmic decisions must have a voice in shaping AI design, deployment, and oversight. Public engagement, stakeholder consultation, and transparency ensure that AI serves the collective interest. Flourishing arises when humans exercise both agency and oversight, embedding fairness and justice into the technological systems that increasingly shape daily life.

Concluding the discussion on bias in code, it is clear that humans shape the ethical and social outcomes of AI. Algorithms do not act independently; they reflect the values, assumptions, and data selected by their creators. Ignoring bias risks perpetuating historical inequities, while conscious design and oversight can mitigate harm and promote fairness. Flourishing arises when humans take responsibility for the ethical deployment of AI, ensuring technology empowers rather than marginalizes.

Bias also underscores the ongoing nature of ethical responsibility. AI systems evolve, data changes, and societal dynamics shift, creating new challenges over time. Humans must maintain vigilance, continuously auditing systems, updating ethical frameworks, and intervening to correct discriminatory patterns. Flourishing depends on active stewardship, combining technical skill, moral reflection, and social awareness to safeguard equity and justice.

Transparency and accountability remain essential. Humans must understand algorithmic logic, scrutinize outcomes, and clarify decision-making processes. Open systems and explainable models enable oversight, fostering trust and ethical alignment. Flourishing emerges when humans can evaluate, guide, and correct AI outputs, ensuring fairness and collective well-being are prioritized over narrow efficiency or profit.

Collaboration and inclusion enhance mitigation of bias. Diverse teams, participatory design, and cross-disciplinary input illuminate hidden inequities and ethical considerations. Human judgment, informed by varied perspectives, ensures AI serves broader societal interests. Flourishing occurs when technology is co-created responsibly, reflecting equity, ethical reflection, and social accountability in every stage of its development and deployment.

Finally, bias in AI is a reminder that ethical responsibility cannot be delegated to machines. Humans must define fairness, enforce accountability, and guide technology with intentionality. Flourishing arises when humans combine technical expertise, ethical vigilance, and social commitment to ensure AI enhances justice, opportunity, and dignity. Technology is a mirror of human choices, and it is up to humanity to reflect the best of its values.

Inspirational Quote:

"*AI reflects the values we code into it; fairness requires humans to design, guide, and hold it accountable.*" — Anonymous

Chapter . 16

Black, White, Brown, and Invisible

Racialized communities experience AI differently based on history and access.

AI does not operate in a societal vacuum. Its development, deployment, and impact are deeply intertwined with historical and structural inequities. Racialized communities often experience AI systems differently, as these systems reflect past injustices and existing power imbalances. From predictive policing to healthcare algorithms, marginalized groups may face disproportionate scrutiny, misrepresentation, or exclusion. Flourishing requires awareness of these disparities and deliberate strategies to ensure equitable access, fair treatment, and inclusive design. Technology amplifies existing realities, and humans must guide it toward justice rather than replication of historical inequities.

Black, white, brown, and Indigenous communities have distinct experiences shaped by social, economic, and political contexts. AI applications can unintentionally reinforce stereotypes, disadvantage certain groups, or fail to represent cultural nuances. Misclassification, underrepresentation, and biased decision-making are common outcomes when systems are trained on non-diverse data. Flourishing emerges when humans actively address these disparities through inclusive datasets, diverse design teams, and culturally informed ethical oversight. Technology must reflect society's values, not only technical efficiency.

Access to AI technologies is also racially and economically stratified. Wealthier and predominantly white communities often benefit first from emerging AI tools, educational resources, and economic opportunities. Meanwhile, communities of color and historically marginalized populations may face barriers to training, adoption, and meaningful participation. Ethical stewardship demands interventions that democratize access, support capacity building, and bridge technological divides. Flourishing occurs when AI empowers all communities equitably, rather than deepening systemic inequalities.

AI also interacts with societal biases in subtle and pervasive ways. Automated hiring, loan approvals, criminal justice algorithms, and social service distribution can all reproduce historical patterns of discrimination. Bias is not always intentional but emerges from embedded data patterns, historical inequities, and societal assumptions. Humans must examine these systems critically, ensuring fairness, accountability, and transparency. Flourishing depends on combining technical solutions with ethical vigilance and social responsibility.

Historical context is essential to understanding how AI affects racialized communities. Systems trained on biased data inherit the inequities of previous generations, amplifying disparities in education, employment, healthcare, and policing. Without intervention, these systems may reinforce cycles of disadvantage, marginalization, and social exclusion. Humans must actively design algorithms to counteract these patterns, incorporating fairness metrics, inclusive data, and oversight mechanisms. Flourishing emerges when AI acknowledges history and is intentionally directed toward equity and justice.

Representation in AI design teams is critical. Diverse teams bring perspectives that reveal hidden biases, challenge assumptions, and guide culturally sensitive applications. Homogeneous teams may overlook subtle inequities or fail to recognize harmful consequences for marginalized communities. Flourishing occurs when humans intentionally foster diversity in design, testing, and deployment, ensuring AI serves the needs and rights of all people, not only those in positions of privilege.

Education and digital literacy also influence racialized impacts. Communities with limited access to AI education and training are less able to engage critically, participate in development, or advocate for ethical practices. This lack of representation reinforces exclusion and systemic inequities. Equitable educational initiatives, mentorship programs, and capacity-building efforts empower marginalized communities to influence AI's direction. Flourishing arises when all communities have the tools and knowledge to shape technology in their own interest.

AI applications in law enforcement, lending, and healthcare exemplify racial disparities. Predictive policing may disproportionately target neighborhoods of color, while credit scoring algorithms may undervalue applicants from historically marginalized groups. Health algorithms trained on biased datasets may fail to recognize disease patterns prevalent in nonwhite populations. Humans must scrutinize, audit, and correct these systems continuously. Flourishing requires vigilance, ethical reflection, and structural accountability to prevent harm and ensure equitable outcomes.

Racialized impacts of AI are compounded by socioeconomic factors. Communities of color often face limited access to technology, funding, and digital infrastructure, restricting their ability to leverage AI for economic or educational advancement. This technological gap reinforces existing inequalities, creating a cycle where those already disadvantaged are further marginalized. Flourishing depends on targeted interventions that increase access, resources, and representation, ensuring AI serves as a tool for opportunity rather than exclusion.

AI also interacts with cultural assumptions embedded in datasets. Facial recognition, natural language processing, and image analysis systems often underperform for nonwhite populations due to skewed or incomplete training data. Errors in identification, translation, or interpretation can have serious consequences, including denial of services, misdiagnosis, or legal injustice. Humans must scrutinize these technologies, adjust datasets, and develop culturally sensitive approaches. Flourishing arises when technology reflects the diversity and complexity of human societies rather than perpetuating homogenized standards.

Bias in AI can shape social narratives. Algorithms that prioritize certain voices, content, or outcomes may reinforce stereotypes or exclude minority perspectives. This is evident in search results, social media feeds, and recommendation systems that influence public perception and social discourse. Humans must critically evaluate how AI amplifies or suppresses specific voices and ensure that representation is fair and accurate. Flourishing occurs when AI is used to elevate inclusive narratives, empower marginalized communities, and foster equitable dialogue.

Intersectionality is also critical. AI may compound disadvantages for individuals who belong to multiple marginalized groups, such as women of color or immigrants in underserved communities. Systems that fail to account for overlapping inequities risk reinforcing multi-layered discrimination. Humans must integrate intersectional awareness into design, testing, and governance of AI. Flourishing emerges when technology acknowledges complex identities and works to mitigate compounded disparities rather than exacerbate them.

Community engagement is crucial to addressing AI's racialized impacts. Those directly affected by algorithmic decisions should have meaningful input into system design, deployment, and oversight. Participatory governance models ensure that AI serves the needs and values of all communities, rather than privileging a select few. Flourishing emerges when humans actively involve diverse populations in shaping technology, fostering equity, trust, and empowerment.

Policy and regulation play an essential role in protecting marginalized communities. Anti-discrimination laws, fairness audits, and ethical guidelines for AI deployment help prevent systemic bias and inequitable outcomes. Regulatory oversight ensures that developers and organizations are held accountable for their algorithms' societal impacts. Flourishing arises when governance mechanisms complement ethical design, ensuring that AI contributes to justice, opportunity, and collective well-being.

Education and capacity-building initiatives are equally important. By providing access to AI literacy, technical training, and decision-making tools, marginalized communities can actively participate in technological development and advocacy. Empowered communities can identify bias, suggest improvements, and co-create systems that serve their interests. Flourishing occurs when humans intentionally cultivate knowledge, agency, and representation across all segments of society.

The invisibility of some communities in data collection and AI decision-making highlights the need for inclusive datasets. Underrepresented groups risk being ignored, misclassified, or misunderstood, leading to systemic neglect. Human responsibility involves curating datasets thoughtfully, ensuring that all voices and experiences are captured accurately. Flourishing depends on inclusivity at the data level, which serves as the foundation for fair, ethical, and equitable AI applications.

Concluding the discussion on racialized impacts, it is clear that AI mirrors societal structures, amplifying both opportunity and inequality. Black, white, brown, Indigenous, and invisible communities experience AI differently based on history, access, and representation. Technology can empower or marginalize depending on human choices in design, deployment, and governance. Flourishing arises when humans intentionally create systems that recognize diversity, honor equity, and ensure inclusion.

AI offers the potential to correct historical injustices—but only if humans actively design for fairness. Algorithms can identify disparities, highlight inequities, and expand opportunity when guided by ethical principles and inclusive data. Conversely, neglecting diversity in design, oversight, and governance risks reproducing the very inequities technology should alleviate. Flourishing depends on ethical stewardship, continuous vigilance, and conscious intervention to ensure AI serves justice and collective well-being.

Community engagement remains essential. Those most affected by AI decisions must have a voice in shaping technology, from policy-making to implementation. Participatory design, consultation, and oversight empower communities to hold AI accountable and influence outcomes. Flourishing emerges when humans prioritize inclusion, dialogue, and shared responsibility, ensuring that AI serves society equitably.

Education, representation, and access form the foundation for ethical AI. By equipping marginalized communities with knowledge, tools, and agency, humans can democratize technology and prevent exclusion. Inclusive design and capacity-building initiatives create systems that reflect the complexity of society and uphold fairness. Flourishing occurs when access, literacy, and empowerment accompany technological advancement.

Finally, ethical reflection is inseparable from technological development. Humans must continuously examine the societal impact of AI, challenge assumptions, and act to prevent harm. AI is not inherently fair; its outcomes reflect human decisions, values, and oversight. Flourishing arises when humans combine technical skill, moral awareness, and social responsibility to create AI that empowers all communities and fosters justice, equity, and dignity.

Inspirational Quote:

"*AI mirrors the society we build; when guided by inclusion and equity, it becomes a force for justice and empowerment.*" — Anonymous

Chapter . 17

AI and the Global South

Emerging economies face both unprecedented opportunity and new exploitation.

AI presents both promise and peril for the Global South. Emerging economies have the potential to leapfrog traditional development stages, improving healthcare, education, agriculture, and infrastructure through AI-driven solutions. At the same time, these regions face risks of exploitation, unequal access, and dependency on technology developed elsewhere. Flourishing depends on balancing opportunity with agency, ensuring that AI serves local needs, builds capacity, and promotes equitable growth. Technology alone is insufficient; human governance, ethical oversight, and strategic planning are essential.

The Global South faces structural challenges that influence AI adoption. Limited infrastructure, inadequate access to data, and scarcity of technical expertise can constrain the benefits of AI deployment. International corporations and institutions often control AI tools, datasets, and resources, creating dependencies and power imbalances. Humans must address these structural inequalities to ensure that AI supports local innovation and empowerment. Flourishing emerges when technology is adapted to local contexts, with communities driving priorities, solutions, and governance.

AI offers significant potential in healthcare, agriculture, and education. Predictive analytics can improve disease prevention and treatment, while machine learning can optimize crop yields and resource allocation. Educational platforms powered by AI can provide personalized learning at scale. However, without local adaptation, these technologies may fail to meet cultural, linguistic, and contextual needs. Flourishing arises when AI is designed collaboratively, reflecting local realities and serving the people it aims to benefit.

Global dynamics also shape AI deployment in the Global South. Access to capital, international partnerships, and intellectual property laws influence who benefits and who is excluded. Emerging economies may struggle to negotiate fair terms or maintain control over technology developed by multinational corporations. Human intervention, policy frameworks, and ethical guidelines are essential to prevent exploitation and ensure equitable distribution of AI benefits. Flourishing depends on sovereignty, agency, and ethical stewardship in adopting AI technologies.

AI adoption in the Global South must navigate ethical and social considerations. Algorithms trained on data from developed countries may not reflect local realities, leading to errors, misdiagnoses, or inequitable outcomes. Cultural, linguistic, and societal differences require careful adaptation of AI systems to ensure relevance and fairness. Humans must actively design, monitor, and evaluate technology to prevent harm and maximize benefit. Flourishing arises when AI respects local contexts, serving communities ethically and effectively.

Dependency on foreign AI providers is a major concern. When multinational corporations control critical infrastructure, data, and algorithms, emerging economies risk losing autonomy over decision-making and innovation. Profit-driven deployment may prioritize external interests over local needs. Humans must advocate for capacity-building, local innovation, and equitable partnerships to maintain agency and sovereignty. Flourishing depends on combining technological adoption with self-determination and ethical governance.

Economic impacts are significant. AI can boost productivity, optimize resource allocation, and create new markets, but it can also exacerbate inequality if benefits are concentrated in elite sectors. Job displacement may occur in sectors reliant on routine labor, disproportionately affecting vulnerable populations. Humans must implement policies that promote fair distribution of AI benefits, workforce retraining, and inclusive growth strategies. Flourishing emerges when technology contributes to economic empowerment rather than widening disparity.

Collaboration and partnerships are key to ethical AI adoption in the Global South. Local governments, universities, NGOs, and businesses must co-create AI systems that reflect societal priorities and ethical standards. International partnerships should emphasize knowledge transfer, local capacity, and equitable participation. Flourishing occurs when AI is not imposed externally but developed in collaboration with communities, ensuring relevance, fairness, and sustainable impact.

AI's role in governance and public services in the Global South is transformative but fraught with risk. Predictive analytics can improve resource allocation, optimize infrastructure, and enhance policy decision-making. However, if systems are poorly designed, biased, or controlled externally, they can entrench corruption, reinforce inequities, or undermine democratic accountability. Humans must ensure transparency, inclusivity, and ethical oversight in AI deployment. Flourishing emerges when AI strengthens governance while respecting the rights and dignity of all citizens.

Data sovereignty is a critical challenge. Many Global South countries lack control over the collection, storage, and use of their own data, leaving it in the hands of multinational corporations or foreign institutions. This imbalance can limit local innovation, skew development priorities, and erode autonomy. Humans must advocate for policies that protect data, encourage local innovation, and establish ethical frameworks for its use. Flourishing occurs when AI development is guided by principles of sovereignty, equity, and ethical responsibility.

AI also presents opportunities for addressing climate and environmental challenges. Machine learning can optimize energy use, monitor deforestation, and predict climate-related risks. Yet without careful management, AI deployment can exacerbate environmental strain, such as energy consumption from data centers or the unequal distribution of benefits. Humans must integrate sustainability, fairness, and local context into AI systems. Flourishing arises when technology serves ecological stewardship as well as social and economic development.

Digital inclusion is essential to prevent widening inequalities. Rural populations, low-income communities, and marginalized groups may lack access to AI tools, internet connectivity, and technical education. Ensuring equitable access requires investment in infrastructure, training programs, and inclusive technologies. Humans must actively bridge the digital divide to empower all communities. Flourishing occurs when AI adoption is accompanied by initiatives that foster inclusion, opportunity, and human dignity.

AI in the Global South also intersects with labor and employment. Automation and intelligent systems can improve efficiency and productivity, but they can also displace workers, particularly in low-skill or routine jobs. Without proactive policies for retraining, social protection, and inclusive growth, AI may deepen economic inequality. Humans must integrate labor considerations into AI strategy, ensuring that technology creates opportunity rather than hardship. Flourishing arises when AI adoption balances economic growth with social equity and human dignity.

Education and skills development are central to realizing AI's potential. Countries in the Global South must invest in technical training, digital literacy, and innovation ecosystems to build local expertise. By fostering local talent, communities can develop, adapt, and govern AI systems rather than remain passive consumers of technology. Flourishing occurs when humans equip themselves with knowledge and agency to guide AI toward societal benefit.

Ethical frameworks must underpin AI deployment. Societal values, cultural norms, and historical contexts should inform the design, implementation, and governance of technology. Neglecting these considerations risks harm, exploitation, or resistance from affected communities. Humans must embed ethical reflection and participatory governance into AI initiatives. Flourishing emerges when technology is aligned with societal values and human rights, ensuring inclusive, responsible, and culturally sensitive application.

Global cooperation can amplify positive outcomes. Partnerships with international organizations, universities, and ethical consortia provide resources, training, and shared knowledge. However, these collaborations must avoid neocolonial dynamics, where foreign interests dominate and local priorities are sidelined. Humans must advocate for equitable, respectful, and reciprocal partnerships. Flourishing occurs when global collaboration empowers local communities and strengthens self-determined development.

Concluding the discussion on AI in the Global South, it is clear that technology is both an opportunity and a responsibility. Emerging economies can harness AI to accelerate development, improve quality of life, and solve pressing societal challenges. At the same time, external control, inequality, and exclusion pose significant risks if local communities are not empowered. Flourishing depends on humans actively shaping AI adoption to serve ethical, inclusive, and locally relevant purposes.

Capacity-building and local expertise are critical for sustainable AI integration. By investing in education, training, and innovation ecosystems, communities gain the agency to develop, govern, and adapt AI technologies. Dependency on external providers diminishes, and technology becomes a vehicle for self-determined progress. Flourishing arises when humans combine knowledge, ethics, and innovation to ensure AI serves local needs and long-term societal goals.

Ethical stewardship must remain central. AI deployment should respect human rights, cultural values, and societal priorities. Participatory governance, transparency, and accountability mechanisms prevent exploitation and ensure that AI benefits are equitably distributed. Flourishing emerges when humans embed ethical reflection and inclusive decision-making into every stage of AI adoption, creating systems that empower rather than marginalize.

Global partnerships must be equitable and collaborative. Knowledge-sharing, joint innovation, and reciprocal engagement strengthen local capacity while avoiding neocolonial dynamics. Humans must advocate for fair collaboration that respects local expertise and priorities. Flourishing occurs when international cooperation amplifies opportunity and empowerment rather than reproducing historical inequities.

Finally, AI in the Global South illustrates a broader truth: technology is a reflection of human choices and values. Flourishing arises when humans intentionally design, govern, and deploy AI to maximize opportunity, equity, and ethical impact. Emerging economies can harness AI as a force for empowerment, but only if guided by vision, responsibility, and a commitment to inclusive development.

Inspirational Quote:

"AI is a mirror of human choice; in the Global South, it can illuminate opportunity or magnify inequality depending on how we guide it." — Anonymous

Chapter . 18

Class in an Automated Economy

Lower-income and middle-class workers face disproportionate disruption.

Automation and AI are reshaping economic landscapes, with profound implications for class structures. Routine, manual, and clerical jobs are increasingly susceptible to displacement, affecting primarily lower-income and middle-class workers. Meanwhile, high-skill, specialized, and technology-oriented roles often remain insulated, creating widening economic divides. Humans must navigate these changes by fostering inclusive policies, skills development, and social safety nets. Flourishing emerges when AI-driven disruption is managed ethically, ensuring opportunity and security across all socioeconomic groups.

The economic divide is magnified by unequal access to education and reskilling opportunities. Workers in lower-income brackets often face barriers to acquiring the skills needed to thrive in an automated economy, including digital literacy, technical training, and lifelong learning resources. Middle-class workers may experience temporary displacement but can pivot if given appropriate support. Flourishing arises when society prioritizes equitable access to education, training, and opportunities, preventing AI from entrenching class disparities.

Automation also influences wealth distribution. Productivity gains from AI often accrue disproportionately to business owners, investors, and those with capital, while workers bear the brunt of job loss or wage stagnation. This dynamic can exacerbate inequality, eroding social mobility and economic stability. Humans must design policies that distribute the benefits of AI fairly, including profit-sharing, tax structures, and investment in public goods. Flourishing occurs when technology enhances collective prosperity rather than concentrating wealth.

Social and psychological consequences of class-based disruption are profound. Job insecurity, reduced income, and diminished social status can lead to stress, anxiety, and decreased life satisfaction. The identity of work as a source of dignity and purpose is challenged, particularly for communities with limited alternatives. Humans must provide social support systems, meaningful opportunities, and pathways for engagement beyond traditional labor. Flourishing depends on fostering resilience, dignity, and agency in the face of economic transformation.

The future of work in an automated economy is uncertain, especially for lower-income and middle-class workers. Many jobs will evolve rather than disappear entirely, but the skills required may shift dramatically. Humans must anticipate these transitions by investing in lifelong learning, upskilling, and adaptable career pathways. Flourishing emerges when society prepares workers to engage meaningfully with new opportunities, minimizing disruption while maximizing human potential.

AI-driven automation can create new industries and markets, but entry barriers often favor those with capital, education, and social connections. Lower-income and middle-class workers may struggle to access these opportunities without deliberate interventions. Policies such as affordable training programs, apprenticeships, and inclusive entrepreneurship initiatives can help bridge the gap. Flourishing occurs when economic opportunity is not reserved for the privileged but is accessible to all social strata.

Inequality is further compounded by the digital divide. Access to high-speed internet, advanced devices, and AI-driven tools is uneven across classes, limiting the ability of marginalized populations to benefit from automation. Humans must invest in infrastructure, technology access, and digital literacy to prevent further marginalization. Flourishing depends on ensuring that technology does not amplify existing disparities but instead creates inclusive pathways for advancement.

The psychological effects of economic disruption cannot be underestimated. Job loss, uncertainty, and downward mobility can undermine self-worth, social cohesion, and community stability. Human flourishing requires not only financial support but also psychological, social, and communal reinforcement. Programs that foster resilience, social networks, and purpose beyond employment can mitigate these effects. AI should be integrated thoughtfully, with attention to both material and emotional well-being.

Labor market transformation in an automated economy is uneven across sectors. Manufacturing, logistics, and administrative jobs are particularly vulnerable to AI and robotics, disproportionately affecting lower-income workers. Meanwhile, sectors requiring creativity, complex problem-solving, or social intelligence may see growth, benefiting those with access to advanced education and training. Humans must recognize these imbalances and implement targeted strategies to support displaced workers. Flourishing arises when AI integration is accompanied by policies that ensure no community is left behind.

Reskilling and upskilling are critical strategies for mitigating class-based disruption. Governments, educational institutions, and private organizations must create accessible programs to equip workers with skills relevant to emerging job markets. These programs should combine technical training with critical thinking, problem-solving, and interpersonal skills. Flourishing occurs when humans proactively invest in lifelong learning, empowering workers to adapt and succeed in a rapidly evolving economy.

Automation also affects social mobility. Jobs that once provided pathways out of poverty may vanish, making it harder for lower-income individuals to achieve economic advancement. This dynamic risks entrenching class divides and limiting opportunity for future generations. Humans must design social safety nets, equitable access to education, and policies that promote upward mobility. Flourishing arises when society consciously safeguards pathways to opportunity, preventing AI from cementing systemic inequality.

The rise of gig and platform economies presents both opportunity and risk. While AI enables flexible work and new income streams, these positions often lack benefits, security, and stability, disproportionately affecting lower-income workers. Humans must create regulations and systems that ensure fair compensation, labor rights, and protections in a rapidly changing job landscape. Flourishing occurs when work structures balance flexibility, opportunity, and social security, empowering rather than exploiting workers.

Economic policy plays a crucial role in shaping outcomes in an automated economy. Progressive taxation, universal basic income, and targeted subsidies can redistribute wealth generated by AI and automation, mitigating inequality. Without deliberate intervention, gains from AI are likely to concentrate in the hands of capital owners and highly skilled workers. Humans must craft policies that balance innovation with social responsibility. Flourishing arises when technology serves collective prosperity rather than reinforcing entrenched class hierarchies.

The psychological dimension of class disruption is equally important. Job displacement, wage stagnation, and reduced social mobility can erode self-esteem, community cohesion, and societal trust. Humans must recognize that economic well-being is intertwined with emotional and social health. Programs that provide support, mentorship, and opportunities for meaningful engagement can help communities navigate these transitions. Flourishing occurs when humans design AI adoption strategies that consider both material and psychological well-being.

Automation also transforms the nature of work itself. Routine tasks become increasingly automated, while roles emphasizing creativity, emotional intelligence, and complex reasoning expand. Lower-income and middle-class workers may need to pivot into these new roles, requiring access to retraining, mentorship, and social support. Flourishing emerges when society facilitates smooth transitions, equipping individuals with the skills and confidence to thrive in evolving labor markets.

Inequities in access to AI tools and digital platforms exacerbate class divides. Those with limited technology access are less able to leverage AI to improve productivity, education, or income. Addressing the digital divide is therefore essential to ensure that automation benefits are distributed broadly. Humans must invest in infrastructure, affordable technology, and digital literacy programs. Flourishing occurs when all communities can participate fully in a technologically advanced economy.

Concluding the discussion on class in an automated economy, it is evident that AI and automation present both opportunity and risk. Lower-income and middle-class workers face disproportionate disruption, while those with capital, education, and digital access often reap the benefits. Humans have the capacity—and the responsibility—to design systems, policies, and programs that mitigate inequality and empower all socioeconomic groups. Flourishing arises when technology serves the collective good rather than deepening existing class divides.

Social safety nets are essential for navigating economic transitions. Universal basic income, retraining programs, and accessible education help workers adapt to shifting job landscapes. Equitable access to these resources ensures that no individual or community is left behind. Flourishing depends on proactive measures that combine economic support, skills development, and opportunity creation in an automated society.

Psychological and social considerations must accompany economic planning. Job loss, wage instability, and social displacement can erode mental health and community cohesion.

Humans must foster resilience, purpose, and social inclusion through education, mentoring, and community engagement. Flourishing occurs when AI integration considers not only economic outcomes but also human well-being, dignity, and social connection.

Ethical deployment of AI is central to preventing exacerbation of class disparities. Transparency, fairness, and accountability in algorithmic systems ensure that automation does not disproportionately harm vulnerable populations. Humans must guide technology with moral and social responsibility, balancing efficiency with equity. Flourishing emerges when AI is a tool for empowerment, inclusion, and opportunity rather than exclusion and inequity.

Finally, class in an automated economy is a test of society's values and foresight. Humans can choose whether technology widens inequality or enhances collective human potential. Flourishing arises when humans combine ethical governance, inclusive policies, and human-centered innovation to ensure AI benefits all members of society, enabling dignity, opportunity, and shared prosperity.

Inspirational Quote:

"*Automation tests our values; the measure of progress is not just efficiency, but whether every member of society can thrive.*" — Anonymous

Chapter . 19

Immigration, Borders, and Digital Labor

AI reshapes migration, work visas, and global competition for talent.

AI is transforming how labor flows across borders, creating new opportunities and challenges for migrants and host countries. Digital platforms, remote work, and algorithm-driven hiring enable talent to participate in global labor markets without physical relocation. This creates unprecedented flexibility but also introduces competition across borders, often favoring those with digital access and technical skills. Humans must navigate these shifts by establishing fair policies, equitable labor protections, and inclusive systems. Flourishing emerges when global labor mobility balances opportunity, equity, and dignity for workers everywhere.

The nature of immigration is evolving in the age of AI. Traditional pathways such as work visas and residency permits may be disrupted by remote employment, automated recruitment, and AI-managed immigration systems. While some workers gain flexibility, others may face increased precarity as algorithms determine eligibility, skill assessment, and hiring decisions. Humans must ensure that technology supports fair treatment, reduces exploitation, and respects human rights. Flourishing occurs when AI strengthens mobility while safeguarding justice and opportunity.

Digital labor platforms are redefining the global workforce. Remote AI-driven work allows companies in developed economies to access talent from emerging markets at lower costs, creating opportunities for income and skills development. However, this also risks creating a new class of digital precariat, with unstable employment, limited benefits, and reduced bargaining power. Humans must develop regulations, ethical standards, and protections to ensure fair compensation and security. Flourishing arises when digital labor benefits workers rather than exploiting them.

AI also shapes migration patterns and economic incentives. Automated systems may prioritize high-demand skills, influencing who can migrate, work remotely, or access opportunities abroad. Workers from marginalized communities or regions with limited technological access may be excluded, perpetuating inequality. Humans must intervene to ensure equitable access, prevent algorithmic discrimination, and support inclusive global labor systems. Flourishing occurs when technology expands opportunity rather than deepening systemic disparities.

AI-driven recruitment is transforming the global labor market, but it raises significant equity concerns. Algorithms can evaluate resumes, screen candidates, and even conduct preliminary interviews, often favoring those with access to training, certifications, and digital literacy. Marginalized or less technologically fluent workers risk being overlooked despite their skills and experience. Humans must ensure that AI hiring systems are designed to recognize diverse qualifications and reduce bias. Flourishing arises when AI enhances opportunity rather than restricting access based on digital fluency or privilege.

Digital labor platforms are creating new forms of remote work, yet they often lack traditional labor protections. Workers may face irregular pay, limited benefits, and little recourse against exploitation. While AI enables flexible work arrangements, it can also commodify human labor in ways that reduce autonomy and security. Humans must advocate for policies, contracts, and oversight that protect workers' rights in these digital spaces.

Flourishing occurs when technological innovation is paired with fairness, dignity, and social responsibility.

Immigration policies themselves are increasingly influenced by AI. Automated visa processing, risk assessments, and eligibility screening aim to improve efficiency but may unintentionally reinforce systemic biases. Applicants from certain countries or backgrounds may be disadvantaged due to incomplete data, biased algorithms, or flawed assumptions. Humans must actively monitor and correct these systems to ensure equitable treatment. Flourishing emerges when technology serves as a tool for inclusion rather than exclusion in global migration systems.

Global competition for talent intensifies as AI makes remote work more viable. Highly skilled workers from anywhere can compete for jobs in global markets, which may benefit those with education and resources but disadvantage those from under-resourced regions. Humans must design equitable access pathways, reskilling initiatives, and mentorship programs to ensure that global labor markets are fair. Flourishing occurs when AI-driven globalization empowers all workers, reducing barriers to opportunity.

AI is also reshaping the skill sets required for global labor mobility. Workers who can adapt to AI-assisted environments, leverage digital platforms, and demonstrate soft skills such as creativity and emotional intelligence gain a competitive advantage. Those lacking access to education, training, or connectivity may be left behind, exacerbating inequality across countries and social classes. Humans must invest in equitable skill development programs and accessible learning platforms. Flourishing arises when technology-driven skill requirements do not become a barrier to opportunity but a pathway for empowerment.

Remote work mediated by AI has the potential to reduce geographic barriers but introduces new forms of digital dependency. Workers may rely on platforms, algorithms, and automated management tools that dictate productivity, monitor performance, and control access to tasks. This creates power asymmetries, where algorithmic oversight can replace human judgment and negotiation. Humans must ensure these systems are transparent, accountable, and designed to empower workers rather than dominate them. Flourishing occurs when AI enhances autonomy, collaboration, and opportunity for all participants.

The rise of AI-assisted gig work also challenges traditional concepts of labor rights and benefits. Unlike conventional employment, digital laborers may have irregular hours, variable pay, and little job security, despite performing essential economic functions. Ethical frameworks and legal protections are necessary to safeguard worker dignity and social stability. Humans must advocate for labor rights, fair compensation, and protections adapted to a digital, AI-mediated economy. Flourishing arises when innovation coexists with social justice and human welfare.

AI also impacts migration by influencing policy decisions in real time. Governments may use predictive models to assess economic need, skill gaps, or potential social impact of immigrants, but these models carry risks of reinforcing biases or overlooking marginalized groups. Humans must balance efficiency with fairness, ensuring that algorithmic decisions are guided by ethical principles and inclusive policy frameworks. Flourishing occurs when AI informs policy without undermining human judgment or perpetuating discrimination.

AI-enabled digital labor platforms can empower marginalized workers by providing access to global markets, yet they also risk exploitation if left unregulated. Payment delays, unclear contracts, and algorithmic task allocation can disproportionately harm vulnerable populations. Humans must establish fair labor standards, oversight mechanisms, and accessible dispute resolution systems. Flourishing arises when technology expands economic opportunity while safeguarding worker rights, dignity, and agency.

Automation and AI can also influence remittances and economic dependency. Migrants performing digital labor often support families in their home countries, creating flows of wealth that can alleviate poverty but also introduce vulnerabilities to algorithmic fluctuations or platform instability. Humans must ensure that digital labor systems provide consistent, fair compensation and economic security. Flourishing occurs when AI-mediated work strengthens economic resilience without creating new forms of precarity.

Policy coordination between countries is essential to address cross-border labor dynamics. Agreements on fair wages, social protections, taxation, and data governance help prevent exploitation and ensure equitable participation. Humans must advocate for international standards that prioritize worker welfare over corporate efficiency alone. Flourishing emerges when AI-driven global labor markets operate transparently, ethically, and inclusively, benefiting workers and communities worldwide.

Ethical design in AI labor platforms is equally important. Algorithms should be auditable, explainable, and capable of detecting bias or unfair practices. Without such safeguards, digital laborers may face discriminatory task allocation, unfair performance evaluation, or exclusion from opportunities. Humans must embed ethics into every stage of platform design, deployment, and monitoring. Flourishing arises when technology amplifies fairness, accountability, and human dignity across borders.

Concluding the discussion on AI, immigration, and digital labor, it is clear that technology reshapes not only where people work, but how, under what conditions, and with what protections. AI can open global opportunities for skilled and digitally literate workers, while simultaneously creating new vulnerabilities for those without access or bargaining power. Humans must guide these transformations with ethical foresight, ensuring that labor remains dignified, fair, and inclusive. Flourishing arises when AI serves as a bridge for opportunity rather than a source of exploitation.

Equitable access to education, training, and digital infrastructure is essential for preparing workers to thrive in AI-mediated labor markets. Marginalized populations must have opportunities to develop skills, participate in global platforms, and secure fair compensation. Humans must advocate for policies and programs that level the playing field and empower communities. Flourishing occurs when technology enhances opportunity across social, economic, and geographic boundaries.

Labor protections must adapt to the realities of remote, AI-driven work. Traditional employment frameworks often fail to account for algorithmically managed tasks, gig arrangements, and cross-border digital labor. Humans must create regulatory structures that ensure fairness, accountability, and security. Flourishing arises when AI-driven work maintains human dignity and aligns with ethical, social, and legal standards.

Global collaboration is necessary to prevent exploitation and promote inclusive opportunity. Countries, corporations, and civil society organizations must coordinate to establish norms, share knowledge, and enforce labor standards. Humans must ensure that the benefits of digital labor are shared broadly, preventing technological dependence from becoming structural disadvantage. Flourishing occurs when AI serves as a tool for equitable, sustainable, and empowering global labor practices.

Finally, human agency remains central. Workers, policymakers, and communities must actively shape AI systems, ensuring they are transparent, ethical, and accountable. Technology is not a neutral arbiter; it reflects the values and choices of those who design and govern it. Flourishing arises when humans retain control, oversight, and responsibility, creating a global labor ecosystem where AI amplifies opportunity, equity, and human dignity.

Inspirational Quote:

"*AI crosses borders, but justice and fairness must travel with it to ensure opportunity for all.*" — Anonymous

Part V — Work, Wealth, and the End of Necessity

Chapter . 20

The Automation of Labor

Many jobs will disappear, but human contribution will not.

Automation and AI are fundamentally reshaping the nature of work. Machines are increasingly capable of performing tasks once thought to require uniquely human skill, from data analysis to pattern recognition and even creative production. This transformation threatens to eliminate certain jobs while creating entirely new opportunities. Humans must adapt by cultivating skills, flexibility, and resilience to remain valuable contributors in an automated world. Flourishing arises when human ingenuity, judgment, and emotional intelligence complement technological efficiency.

Not all jobs are equally vulnerable to automation. Roles involving repetitive, routine, or predictable tasks are at the highest risk, while positions requiring creativity, empathy, and complex problem-solving are more resilient. Humans must identify areas where uniquely human strengths—like ethical reasoning, emotional labor, and nuanced judgment—are indispensable. Flourishing occurs when society values these contributions and ensures humans and machines work in tandem.

Economic impacts of automation are uneven. While businesses may benefit from reduced labor costs and increased productivity, workers may experience displacement, wage stagnation, or uncertainty. Lower-income and middle-class employees often bear the brunt of these changes, while capital owners capture the majority of gains. Humans must develop policies, social safety nets, and retraining programs to distribute the benefits of automation equitably. Flourishing arises when economic gains are shared and workers are empowered to adapt and thrive.

Automation also changes the social meaning of work. For many, identity, purpose, and social connection are closely tied to employment. As machines assume routine functions, humans must redefine what it means to contribute meaningfully to society. Flourishing depends on cultivating purpose beyond traditional labor, embracing creativity, learning, and community engagement as essential forms of human value. Technology should augment human purpose rather than replace it.

The pace of automation is accelerating, driven by advances in machine learning, robotics, and AI-driven analytics. Industries that once relied heavily on human labor are rapidly integrating intelligent systems to optimize efficiency and reduce costs. This shift creates both opportunities and challenges: businesses benefit from productivity gains, while workers face uncertainty about job security and career longevity. Humans must embrace adaptability, continuous learning, and innovative thinking to remain relevant. Flourishing occurs when individuals and societies align human potential with technological progress.

Automation does not eliminate the need for human oversight. Even the most sophisticated machines require monitoring, interpretation, and intervention when unexpected situations arise. Humans provide judgment, creativity, and ethical reasoning that machines cannot replicate. Flourishing arises when humans recognize the complementary nature of human and machine intelligence, designing systems that leverage the strengths of both.

Certain sectors experience automation differently. Manufacturing, logistics, and clerical work are highly susceptible, while healthcare, education, and creative industries face slower adoption.

Humans must anticipate these sector-specific impacts, tailoring training, policy, and workforce development to meet emerging needs. Flourishing occurs when strategies are proactive, targeted, and inclusive, ensuring that no worker is left behind in the transition.

Automation also reshapes organizational hierarchies. Decision-making may become more data-driven, relying on algorithmic insights rather than human intuition alone. While this can improve efficiency and reduce error, it also risks diminishing human agency, creativity, and leadership. Humans must assert their role in guiding technology, ensuring that organizational structures balance machine efficiency with human judgment. Flourishing emerges when humans retain control over the systems that increasingly shape work and society.

Automation is not merely a technological shift; it is a social transformation. Work provides identity, purpose, and connection, and as AI assumes routine functions, humans must redefine their roles in society. Many will find opportunity in creative problem-solving, emotional intelligence, and innovation—areas where machines cannot replicate human nuance. Flourishing occurs when societies recognize and celebrate uniquely human contributions, ensuring that meaning and purpose extend beyond tasks performed.

Job displacement is inevitable in certain sectors, and the pace of change may outstrip traditional social safety nets. Governments, organizations, and communities must proactively implement strategies to support affected workers, including retraining programs, unemployment protections, and educational initiatives. Humans must create pathways for transition, not simply react to disruption. Flourishing arises when systems are designed to guide people through change, preserving dignity, opportunity, and hope.

The ethical deployment of automation is paramount. Decisions about which jobs to automate, which processes to optimize, and which workers to support reflect societal priorities and moral responsibility. Humans must ensure that automation is applied fairly, transparently, and with consideration for human well-being. Flourishing emerges when ethical reasoning shapes technological development, preventing exploitation or inequity in the workplace.

Collaboration between humans and machines is a defining feature of the automated economy. Rather than viewing AI as a competitor, humans can leverage its strengths to enhance productivity, creativity, and insight. This requires training, adaptation, and mindset shifts that prioritize synergy over replacement. Flourishing occurs when humans and machines complement each other, creating outcomes that neither could achieve alone.

Automation also has profound implications for economic inequality. As machines assume routine work, wealth may concentrate among those who own and control technology, while displaced workers face insecurity and reduced opportunity. Humans must develop policies that ensure fair distribution of automation's benefits, including social safety nets, progressive taxation, and access to education. Flourishing occurs when society mitigates inequality and ensures that economic gains from AI are shared broadly, rather than concentrated in the hands of a few.

The psychological effects of automation are equally important. Job displacement, uncertainty, and changes to identity can lead to stress, anxiety, and social disruption. Humans must prioritize mental health, social support, and purpose-driven opportunities to navigate this transformation. Flourishing arises when societies integrate psychological well-being into the design of automation policies and workforce strategies, acknowledging that humans are more than economic units.

Work itself may evolve into a collaborative endeavor between humans and AI. Machines can handle repetitive, analytical, or data-intensive tasks, freeing humans to focus on creative, strategic, and interpersonal work. This shift emphasizes uniquely human capacities, reinforcing the value of judgment, empathy, and innovation. Flourishing occurs when human work complements AI capabilities, creating synergy that enhances productivity and satisfaction.

Automation also requires a rethink of organizational structures. Decision-making processes, hierarchy, and management strategies must adapt to an environment where AI contributes significantly to operational insight. Humans must balance efficiency and accountability with human judgment and ethical responsibility. Flourishing emerges when organizations integrate technology thoughtfully, preserving human agency while optimizing performance.

Concluding the discussion on the automation of labor, it is clear that work is changing in fundamental ways, but human contribution remains essential. Machines can handle repetitive, analytical, and data-driven tasks, yet uniquely human qualities—creativity, empathy, ethical reasoning, and judgment—cannot be replicated. Humans must embrace adaptability, continuous learning, and collaboration with technology to maintain relevance and purpose. Flourishing arises when humans leverage AI as a partner rather than a replacement, creating opportunities for innovation and meaningful engagement.

Economic adaptation is crucial for mitigating disruption. Policies such as reskilling programs, social safety nets, and inclusive economic frameworks help distribute the benefits of automation fairly. Humans must ensure that technological gains do not exacerbate inequality but instead empower communities across socioeconomic levels. Flourishing occurs when economic structures align with ethical principles, supporting both prosperity and human dignity.

Psychological and social well-being must be integrated into automation strategies. Job displacement and the erosion of traditional work roles can undermine identity, purpose, and social cohesion. Humans must cultivate environments that promote resilience, mentorship, and pathways to meaningful contribution beyond employment. Flourishing emerges when societies prioritize both economic and emotional health in the era of automation.

Ethics and governance guide how automation impacts society. Decisions about which jobs to automate, how to implement AI, and how to support displaced workers reflect values, priorities, and responsibility. Humans must ensure that automation enhances opportunity, fairness, and human flourishing rather than solely maximizing efficiency. Flourishing arises when technology is guided by moral consideration and human-centered design.

Finally, the automation of labor challenges humans to redefine the meaning of work and contribution. Humans have the unique capacity to create, empathize, and innovate in ways that machines cannot replicate. Flourishing emerges when society embraces this potential, designing systems that amplify human ingenuity while leveraging AI to handle tasks that free humans to focus on higher-order contributions.

Inspirational Quote:

"*Automation removes routine, but it cannot replace the creativity, judgment, and heart that make us human.*" — Anonymous

Chapter . 21

When Work Becomes Optional

AI raises the realistic possibility of a post-work society.

Advances in AI and automation are bringing humanity closer to a world where traditional work is no longer necessary for survival. Machines increasingly perform tasks once thought essential, from manufacturing and logistics to analysis and even creative production. This shift raises profound questions about purpose, identity, and the meaning of contribution. Humans must explore how to thrive in a society where labor is optional, reimagining productivity, creativity, and social engagement. Flourishing emerges when human value is disentangled from employment, allowing for freedom and self-actualization.

The possibility of optional work challenges long-standing cultural norms. For centuries, work has been equated with worth, dignity, and social status. As AI takes on more labor, humans must cultivate new frameworks to measure contribution, success, and fulfillment beyond economic output. Flourishing arises when society embraces alternative forms of meaningful engagement, such as creativity, caregiving, learning, and community service.

Optional work also demands a rethinking of economic systems. Without labor as the primary means of income, mechanisms like universal basic income, wealth redistribution, and social support become crucial to ensure survival and equality. Humans must design policies that provide security while encouraging engagement and purpose. Flourishing occurs when economic stability is decoupled from employment, enabling individuals to pursue activities that nurture growth, creativity, and community.

The psychological impact of optional work is complex. Freedom from necessity can liberate, but it can also create existential uncertainty and a sense of purposelessness. Humans must cultivate meaning through self-directed projects, learning, artistic expression, and social contribution. Flourishing emerges when individuals are empowered to define their own purpose rather than relying on externally imposed roles or economic pressures.

Optional work also challenges traditional educational systems. Schools and training programs have historically prepared individuals to enter the workforce, but a post-work society requires new approaches focused on creativity, critical thinking, ethical reasoning, and personal development. Humans must reimagine learning as a lifelong pursuit aimed at cultivating curiosity, resilience, and the ability to contribute meaningfully in non-economic ways. Flourishing occurs when education nurtures purpose, passion, and adaptability rather than solely preparing students for jobs.

Social structures will also evolve in a society where work is optional. Community engagement, volunteerism, and caregiving may become primary avenues for contribution, shifting the emphasis from income to impact. Humans must create systems that recognize and reward these forms of labor, ensuring that all contributions are valued equally. Flourishing arises when societal recognition extends beyond paid work, affirming the dignity and worth of diverse forms of human effort.

The role of technology itself must be reconsidered. AI can provide not only labor replacement but also tools to enhance learning, creativity, and collaboration. Humans must use these technologies intentionally to expand opportunity, reduce inequality, and empower personal growth. Flourishing emerges when AI supports human exploration, innovation, and connection rather than solely optimizing productivity or profit.

Economic inequality remains a central challenge. Even in a world where work is optional, access to resources, technology, and support systems may determine who can fully benefit from freedom. Humans must ensure that policies, wealth distribution, and infrastructure provide equitable access, preventing a new form of social stratification. Flourishing occurs when society designs systems that enable all individuals to thrive, regardless of prior economic status or background.

Optional work also transforms how society measures success. Traditional metrics such as income, job title, and career progression may no longer serve as the primary indicators of human achievement. Humans will need to cultivate new measures of value, including contributions to community, artistic creation, knowledge sharing, and emotional support. Flourishing arises when society recognizes diverse forms of meaningful contribution and reframes what it means to live a successful life.

Mental health and well-being become central in a post-work world. Freed from the necessity of labor, individuals may struggle with direction, motivation, and purpose. Humans must intentionally cultivate practices, routines, and projects that provide structure, growth, and satisfaction. Flourishing occurs when psychological well-being is prioritized alongside material security, enabling individuals to thrive emotionally, socially, and intellectually.

Optional work encourages a reconsideration of human creativity. With survival no longer dependent on employment, people have greater freedom to explore arts, sciences, and innovation. AI can act as a collaborator, expanding human capability while leaving space for original thought. Humans must embrace curiosity, experimentation, and cross-disciplinary exploration to make the most of this newfound freedom. Flourishing emerges when creativity is nurtured and celebrated as a core human endeavor.

Community dynamics will also shift. Humans may invest more time in mentorship, caregiving, and social cohesion, building networks of mutual support and collaboration. Without work as the primary organizing principle, society can focus on collective flourishing and shared purpose. Humans must cultivate empathy, cooperation, and civic engagement to sustain strong, resilient communities. Flourishing occurs when freedom from labor enables deeper social connection and contribution.

The emergence of optional work also impacts intergenerational dynamics. Younger generations may grow up in a society where economic survival is decoupled from labor, altering attitudes toward responsibility, ambition, and contribution. Humans must intentionally guide education, mentorship, and cultural norms to ensure that freedom from work fosters growth, ethics, and engagement rather than complacency or entitlement. Flourishing arises when each generation learns to balance autonomy with purpose, curiosity, and community contribution.

Economic systems must adapt to sustain a post-work society. Without traditional employment driving income, alternative models such as universal basic income, collective ownership, and community wealth-building become essential. Humans must design policies that maintain equitable access to resources while encouraging creativity, innovation, and social participation. Flourishing occurs when society provides security without undermining motivation, autonomy, or personal development.

The role of leisure transforms in a post-work world. Activities that were once secondary or restorative may now occupy central importance, including learning, artistic expression, volunteering, and community building. Humans must redefine how time is valued and organized, creating opportunities for growth, engagement, and fulfillment beyond economic necessity. Flourishing emerges when leisure becomes a space for exploration, creativity, and meaningful connection.

Technology can amplify the benefits of optional work. AI-driven platforms can facilitate education, creative collaboration, and social participation on an unprecedented scale. Humans must intentionally design these tools to expand access, encourage inclusion, and empower individuals to pursue personal and collective growth. Flourishing arises when technology enhances human potential and social cohesion rather than solely focusing on efficiency or profit.

Concluding the discussion on optional work, it is clear that AI brings humanity closer to a post-labor society where survival no longer depends on traditional employment. This transformation offers both opportunity and responsibility: humans must redefine purpose, value, and contribution beyond economic necessity. Flourishing arises when individuals and societies embrace creativity, learning, relationships, and civic engagement as primary measures of meaning and success.

Economic models must evolve to ensure security and equity. Universal basic income, collective ownership, and social safety nets provide the foundation for a society in which labor is optional. Humans must balance stability with freedom, enabling exploration, innovation, and personal growth. Flourishing occurs when economic systems support human well-being without tethering dignity and opportunity to employment.

Psychological and social support structures are essential. Freedom from work may bring liberation, but it also poses challenges to identity, routine, and motivation. Humans must cultivate practices, communities, and educational frameworks that foster purpose, engagement, and mental well-being. Flourishing arises when humans use freedom not for idleness but for growth, creativity, and contribution.

Technology, particularly AI, is a powerful enabler in a post-work society. It can augment learning, collaboration, and social connection while handling tasks once required for survival. Humans must intentionally design AI to empower, rather than dominate, ensuring that machines serve as partners in human flourishing. Flourishing emerges when technology enhances autonomy, creativity, and collective well-being.

Finally, the shift toward optional work is a profound cultural and philosophical moment. Humans have the opportunity to rethink what it means to live a meaningful life, how to contribute to society, and how to nurture communities. Flourishing arises when individuals embrace curiosity, creativity, and connection, using the freedom afforded by AI to cultivate a richer, more intentional existence.

Inspirational Quote:

"*When work is no longer required, the human spirit is free to explore, create, and connect in ways that define true flourishing*."
— Anonymous

Chapter . 22

Universal Basic Income and Economic Dignity

New economic models attempt to separate survival from employment.

Universal Basic Income (UBI) is emerging as a potential solution to the economic disruption caused by AI and automation. By providing a guaranteed income to all individuals, society can decouple survival from employment, allowing humans to focus on creativity, learning, and community engagement. UBI challenges traditional assumptions about work, value, and economic responsibility, presenting an opportunity to reimagine how humans thrive. Flourishing arises when financial security empowers individuals to pursue purpose, innovation, and well-being beyond labor.

The philosophical foundation of UBI is grounded in human dignity. Income should not be contingent upon performing work that machines can increasingly do more efficiently. Humans possess intrinsic value, and society has a moral responsibility to ensure basic security, regardless of labor participation. Flourishing occurs when economic systems affirm dignity, autonomy, and opportunity for every individual, freeing them to pursue meaningful endeavors.

UBI also offers a pathway to reduce inequality and economic precarity. Automation disproportionately affects lower-income and middle-class workers, creating disparities in access to resources and opportunities. By guaranteeing a baseline income, society can buffer against displacement, provide stability, and enable upward mobility. Humans must carefully design UBI systems to ensure fairness, accessibility, and sustainability. Flourishing arises when financial security empowers all members of society, not just the privileged.

Critics of UBI often raise concerns about motivation, inflation, and feasibility. While valid, these challenges can be addressed through careful policy design, phased implementation, and complementary measures such as skills development and social support. Humans must experiment, learn, and adapt policies to balance financial security with societal productivity and engagement. Flourishing occurs when economic innovation is coupled with pragmatic governance and ethical responsibility.

Implementing UBI requires careful consideration of economic sustainability. Funding can come from taxation, wealth redistribution, automation dividends, or public investments in AI-driven productivity gains. Humans must balance fiscal responsibility with the moral imperative to provide security for all. Flourishing occurs when economic policies are structured to ensure long-term stability while empowering individuals to pursue meaningful lives without fear of destitution.

UBI also has the potential to stimulate innovation and entrepreneurship. When survival is guaranteed, individuals are free to take risks, start businesses, and explore unconventional ideas. This freedom can unleash creativity and drive economic diversification, creating a more dynamic society. Humans must design systems that support experimentation and mitigate potential economic pitfalls. Flourishing arises when individuals have both the security and freedom to innovate, contribute, and grow.

Social dynamics may shift significantly under UBI. The traditional link between work and social status could diminish, reducing stigma associated with unemployment or low-income positions. Humans must cultivate cultural norms that value contribution, character, and creativity rather than solely financial or occupational achievement. Flourishing occurs when dignity is preserved and all forms of meaningful human activity are recognized and celebrated.

UBI also provides a framework for addressing inequities caused by automation. Lower-income and middle-class workers, often most affected by job displacement, gain financial stability and access to opportunity. Humans must complement UBI with education, training, and community support to ensure that all individuals can thrive in a rapidly changing economy. Flourishing emerges when economic safety nets and empowerment strategies work together to create equitable outcomes.

UBI challenges the traditional relationship between labor and self-worth. Historically, employment has been a primary source of identity, social status, and purpose. With basic financial security guaranteed, humans can explore contributions beyond paid work, such as caregiving, volunteering, and artistic pursuits. Flourishing occurs when dignity and value are recognized independently of economic output, allowing human potential to expand in new directions.

The psychological benefits of UBI are profound. Reduced financial stress can improve mental health, enhance creativity, and strengthen family and community bonds. Individuals may experience greater freedom to pursue passions, education, and meaningful relationships. Humans must support complementary social structures to maximize these benefits, including access to counseling, mentorship, and collaborative networks. Flourishing emerges when economic security enables emotional, intellectual, and social well-being.

UBI also has implications for gender equity. Historically, caregiving and domestic labor—roles often performed by women—have been undervalued economically. By decoupling income from traditional employment, UBI recognizes and supports these contributions. Humans must ensure that UBI policies are implemented inclusively, addressing historical inequities and empowering all caregivers. Flourishing occurs when society values and supports diverse forms of work, including unpaid but essential contributions.

The environmental impact of UBI should not be overlooked. With reduced pressure to participate in survival-driven labor, humans may make more sustainable choices regarding consumption, commuting, and production. AI-driven planning combined with financial security can encourage ecological responsibility. Flourishing arises when humans leverage economic freedom to make choices that benefit both people and the planet, fostering a sustainable and thriving society.

UBI also redefines the relationship between society and work. In a post-labor framework, humans may engage in activities previously undervalued or unpaid, such as mentoring, creative endeavors, and community organizing. AI can assist by automating routine tasks, providing analytical support, and enhancing productivity in these pursuits. Flourishing arises when humans leverage both freedom and technology to contribute meaningfully in ways that transcend traditional employment.

The implementation of UBI requires careful attention to policy design and equity. Flat distribution may not address regional cost-of-living differences or existing disparities, so adaptive systems and complementary supports are essential. Humans must ensure that UBI reaches all populations equitably, preventing unintended consequences or the reinforcement of systemic inequities. Flourishing occurs when economic safety nets are designed thoughtfully, maximizing inclusivity and empowerment.

UBI can also foster creativity and lifelong learning. Freed from financial pressure, humans can invest time in education, personal growth, and skill development. This shift transforms society from one focused on survival-driven productivity to one that values intellectual exploration and innovation. Flourishing emerges when individuals are encouraged to develop talents, pursue curiosity, and engage with knowledge for its own sake.

The social impact of UBI extends to relationships and community cohesion. With basic financial needs met, people can devote energy to caregiving, volunteerism, and collaboration, strengthening social bonds and civic engagement. Humans must cultivate cultural norms that celebrate these contributions as integral to societal flourishing. Flourishing arises when economic security enables humans to prioritize connection, empathy, and collective well-being over survival pressures.

Concluding the discussion on Universal Basic Income and economic dignity, it is clear that decoupling survival from employment is both a challenge and an opportunity. AI and automation make it increasingly feasible to provide financial security for all, allowing humans to explore purpose, creativity, and contribution beyond traditional work. Flourishing arises when society recognizes human value independently of labor and invests in structures that support opportunity, well-being, and engagement.

UBI empowers individuals to pursue passions, learning, and social impact without the constant pressure of economic survival. Freed from the constraints of necessity, humans can focus on creative endeavors, caregiving, and community engagement. Humans must design complementary policies—education, mental health support, and infrastructure—that maximize the potential of economic freedom. Flourishing occurs when opportunity, dignity, and creativity are nurtured alongside financial security.

Equity is central to the success of UBI. Without careful design, such programs could reinforce existing inequalities or leave marginalized groups behind. Humans must ensure inclusivity, accessibility, and fairness in distribution and implementation. Flourishing emerges when society invests in economic systems that empower everyone, regardless of background or circumstance.

Psychological and social well-being are critical. Financial security reduces stress, increases freedom, and supports mental health, allowing humans to engage more meaningfully with society. Humans must complement UBI with programs that foster social cohesion, mentorship, and personal development. Flourishing arises when individuals have both the resources and support to live purposeful, connected, and fulfilling lives.

Finally, UBI challenges humanity to redefine work, purpose, and societal contribution. Humans have the opportunity to shape a society where creativity, empathy, and collaboration are valued over survival-driven labor. Flourishing occurs when AI, economic innovation, and human-centered policy intersect to create a world where dignity, opportunity, and meaning are universally accessible.

Inspirational Quote:

"*Economic freedom is the soil in which human creativity, purpose, and dignity can truly grow.*" — Anonymous

Chapter . 23

Purpose Without Paychecks

Humans must redefine meaning beyond traditional labor.

As AI and automation increasingly replace routine and even skilled work, humans are challenged to find purpose independent of paychecks. For centuries, employment has been a primary source of identity, social status, and daily structure. The decoupling of income from work invites a profound cultural and personal shift: people must cultivate meaning through creativity, relationships, and community contribution. Flourishing arises when humans redefine their lives around values, passions, and impact rather than economic necessity.

Purpose without paychecks emphasizes intrinsic motivation. Activities such as learning, volunteering, caregiving, mentoring, and artistic creation gain prominence as essential contributions to society. Humans must recognize that meaningful work is not defined solely by compensation but by the difference it makes in the lives of others and in personal growth. Flourishing occurs when individuals value contribution, impact, and engagement over external rewards.

Optional work also reshapes the social fabric. Communities may increasingly rely on collaboration, civic participation, and mutual support rather than purely economic interdependence. Humans must foster networks, cultural norms, and systems that celebrate non-paid contributions, ensuring that identity and respect are not tied to employment. Flourishing arises when society honors human effort in all its forms, regardless of monetary reward.

The psychological dimension is central. Humans derive purpose from challenge, growth, and contribution. Without paychecks, individuals must identify avenues to exercise agency, responsibility, and creativity. Flourishing occurs when humans intentionally design personal routines, goals, and projects that align with passions and values, providing fulfillment beyond survival-driven labor.

Purpose without paychecks also requires rethinking education and lifelong learning. Traditional schooling often prepares individuals for employment, emphasizing measurable skills tied to jobs. In a post-labor or partially automated world, humans must cultivate curiosity, critical thinking, creativity, and emotional intelligence. Flourishing occurs when education encourages exploration, resilience, and personal growth, equipping individuals to find purpose beyond earning a living.

Community engagement becomes a vital avenue for meaningful contribution. Volunteering, mentoring, and civic involvement allow humans to create positive impact while building relationships and social cohesion. Without the necessity of paid work, humans can dedicate more time and energy to collaborative efforts that strengthen society. Flourishing arises when individuals actively participate in communities, recognizing that purpose can be shared, relational, and generative.

Financial independence from labor also fosters entrepreneurial and experimental pursuits. Freed from the pressure to survive, individuals can explore creative projects, start ventures, or engage in research and innovation without the constraints of immediate financial return. Humans must embrace risk-taking, curiosity, and experimentation to unlock new forms of personal and societal value. Flourishing occurs when security enables exploration, innovation, and the pursuit of passions.

Optional work encourages reflection on personal values and ethics. When income is not the driving force, decisions are guided more by meaning, purpose, and long-term impact. Humans must cultivate self-awareness, moral reasoning, and reflective practices to navigate choices and contribute responsibly. Flourishing emerges when individuals align actions with principles, fostering integrity, empathy, and thoughtful engagement with the world.

Purpose without paychecks also reshapes the nature of career planning. Traditional trajectories, focused on climbing ladders or maximizing income, may no longer define success. Humans can pursue portfolios of meaningful activities, combining creative work, civic engagement, mentorship, and exploration. Flourishing arises when individuals embrace diverse forms of contribution, valuing impact and growth over titles and salaries.

The role of technology in enabling purposeful living becomes increasingly important. AI can assist in automating mundane tasks, providing personalized learning opportunities, and facilitating collaboration across distances. Humans must engage intentionally with technology to enhance their passions and initiatives rather than allowing it to dictate focus or attention. Flourishing occurs when technology amplifies human potential, freeing time for creativity, connection, and exploration.

Cultural norms and societal values must evolve to support purpose-driven lives. Recognition, respect, and social status should not be solely tied to paid work but to contribution, innovation, and service. Humans must foster inclusive systems that honor diverse forms of achievement and ensure that everyone has the opportunity to participate meaningfully. Flourishing emerges when societies celebrate a broad spectrum of human endeavor, from caregiving to artistic creation to civic engagement.

The psychological aspect of living without paycheck-driven purpose requires attention. Humans may face existential questions or uncertainty about self-worth when disconnected from traditional work roles. Structured opportunities, mentorship, community involvement, and creative outlets are essential to maintain mental health and personal fulfillment. Flourishing occurs when individuals actively cultivate purpose through engagement, reflection, and meaningful action.

Purpose without paychecks encourages exploration of passions previously sidelined by economic pressures. Individuals may devote time to artistic pursuits, scientific inquiry, or social innovation that were once considered luxuries. Humans must embrace experimentation and the freedom to pursue activities that resonate with personal values and curiosity. Flourishing arises when individuals are empowered to follow interests that cultivate growth, creativity, and meaningful impact.

Collaborative endeavors gain prominence in a post-labor context. Humans can engage more deeply in communities, cooperative projects, and collective problem-solving without the constraints of earning a living. AI can facilitate coordination, provide analytical support, and connect individuals with shared goals. Flourishing occurs when technology and social collaboration work hand in hand to amplify human creativity, empathy, and problem-solving capacity.

Freedom from paycheck-driven work also changes the meaning of productivity. Rather than measuring value by output or financial reward, humans can focus on impact, learning, and contribution. Individuals may combine diverse activities—mentorship, volunteering, personal projects—into meaningful portfolios of engagement. Flourishing emerges when society recognizes and values diverse forms of contribution that enhance human well-being and community cohesion.

Education and skill development are central to supporting purpose without paychecks. Lifelong learning enables individuals to adapt, innovate, and engage in meaningful projects throughout life. Humans must cultivate curiosity, critical thinking, and interdisciplinary skills to navigate a world where opportunity is abundant but structure is self-directed. Flourishing arises when knowledge and growth are pursued for their own sake, not merely as a means to income.

Concluding the discussion on purpose without paychecks, it is evident that human meaning and fulfillment can transcend traditional work. AI and automation free individuals from survival-driven labor, offering the opportunity to explore creativity, relationships, learning, and social impact. Flourishing arises when humans cultivate purpose through intentional action, curiosity, and connection rather than financial necessity alone.

Optional work and UBI create the structural foundation, but individual engagement completes the equation. Humans must actively design lives around passions, values, and contributions that enrich themselves and others. Flourishing occurs when society nurtures environments that support experimentation, creativity, and service, empowering individuals to pursue meaningful endeavors.

Cultural and societal norms must adapt to honor diverse contributions. Status, recognition, and respect should reflect ethical, creative, and social impact rather than purely economic achievement. Humans must cultivate inclusive frameworks that celebrate caregiving, mentorship, artistic creation, and community building as essential forms of value. Flourishing emerges when society affirms all meaningful efforts, ensuring dignity and purpose for everyone.

Psychological well-being is central to living purposefully without paychecks. Humans require structure, community, reflection, and support to navigate freedom from economic necessity successfully. Flourishing occurs when individuals develop routines, relationships, and projects that foster growth, resilience, and fulfillment, ensuring that freedom leads to enrichment rather than stagnation.

Finally, this transformation invites humanity to reconsider the essence of a meaningful life. With survival decoupled from work, humans are called to pursue curiosity, creativity, empathy, and ethical engagement. Flourishing arises when freedom from labor empowers humans to live intentionally, contribute generously, and cultivate both personal and collective well-being.

Inspirational Quote:

"When income is guaranteed, the soul is free to create, connect, and contribute in ways that define true purpose." — Anonymous

Chapter . 24

The Psychological Impact of Jobless Abundance

Freedom from work can liberate or destabilize the human psyche.

The rise of AI and automation brings humanity closer to a world of jobless abundance, where survival is increasingly decoupled from labor. While this freedom presents enormous opportunity, it also introduces complex psychological challenges. Humans may experience uncertainty, boredom, loss of identity, or existential anxiety when traditional work no longer structures daily life. Flourishing arises when individuals cultivate purpose, resilience, and mental well-being to navigate this new landscape.

Jobless abundance can disrupt long-established social and personal identities. Many derive self-worth and social status from their professional roles, and the loss of these structures may cause feelings of inadequacy or disorientation. Humans must develop new frameworks for self-definition based on creativity, relationships, contribution, and personal growth. Flourishing occurs when individuals build identity from intrinsic values rather than external markers of success.

Freedom from economic necessity can create both liberation and pressure. Some may feel empowered to explore passions and learning, while others may struggle with decision fatigue or lack of direction. Humans must cultivate discipline, curiosity, and reflective practices to channel freedom productively. Flourishing arises when autonomy is paired with purpose, enabling individuals to engage meaningfully with life and society.

The psychological effects of abundance are also influenced by social and cultural context. Societies that continue to equate work with value may exacerbate stress, guilt, or alienation in those no longer employed. Humans must foster cultural norms that recognize diverse forms of contribution, including caregiving, artistic creation, volunteering, and mentorship. Flourishing occurs when individuals feel valued and connected, regardless of employment status.

The transition to jobless abundance requires a shift in daily routines and time management. Without the structure of traditional work, humans may struggle to organize their days effectively, leading to stress or a sense of aimlessness. Intentional practices such as goal-setting, creative projects, learning, and community involvement become essential. Flourishing arises when individuals create meaningful rhythms that balance productivity, exploration, and rest.

Social relationships play a critical role in navigating jobless abundance. Humans often connect through work environments, and removing this central point of interaction can create isolation. Building networks, participating in collaborative projects, and nurturing friendships and family bonds become more important than ever. Flourishing occurs when humans cultivate strong, supportive communities that reinforce connection, empathy, and shared purpose.

Mental health support is crucial in a society where work is optional or scarce. Anxiety, depression, and existential stress may increase without the familiar anchors of employment and routine. Humans must provide accessible counseling, mentorship, and community-based resources to support well-being. Flourishing emerges when societies actively integrate mental health care into the fabric of daily life, recognizing that psychological stability is essential for purposeful engagement.

Jobless abundance also opens opportunities for self-actualization. Freed from survival pressures, humans can pursue intellectual, artistic, and social endeavors that were previously constrained by necessity. AI can assist by providing tools for creativity, knowledge exploration, and collaboration. Humans must embrace these opportunities intentionally to cultivate growth and fulfillment. Flourishing occurs when individuals align freedom with curiosity, creativity, and meaningful contribution.

The removal of work as a necessity challenges humans to redefine achievement and success. Traditional markers such as job titles, promotions, and salaries lose relevance, and individuals must cultivate alternative measures of accomplishment. These may include creativity, learning, contribution to others, and community engagement. Flourishing arises when humans value intrinsic fulfillment and societal impact over external validation tied to employment.

Identity formation becomes a central concern in a world of jobless abundance. Without work to structure daily life and social identity, humans may face existential questions about purpose, worth, and direction. Cultivating self-awareness, reflection, and intentional living practices helps individuals navigate this transition. Flourishing occurs when humans construct identity around values, skills, relationships, and contributions rather than occupational labels.

Cultural adaptation is equally important. Societies that equate value with employment may inadvertently create shame or alienation for those not participating in traditional work. Humans must collectively redefine cultural norms to celebrate diverse forms of contribution, including caregiving, mentoring, volunteerism, and artistic expression. Flourishing emerges when society affirms that every human has value, regardless of employment status.

AI and automation can help humans manage the psychological challenges of abundance. Personalized learning, creativity tools, virtual collaboration platforms, and mental health resources can support personal growth and engagement. Humans must ensure that technology amplifies human agency and empowerment, rather than fostering passivity or dependency. Flourishing arises when AI enhances autonomy, creativity, and purposeful participation in society.

Jobless abundance also highlights the importance of community and shared purpose. Humans may no longer rely on workplaces for social interaction, collaboration, or a sense of belonging. Community networks, volunteer projects, creative collectives, and civic engagement become central to mental health and identity. Flourishing arises when individuals participate in shared efforts that create connection, meaning, and collective growth.

Structured routines and goal-setting are essential to prevent drift and stagnation. Freed from work obligations, humans may struggle with time management and motivation. Developing personal schedules, learning plans, and creative projects provides structure and purpose. Flourishing occurs when individuals intentionally design their days to balance rest, exploration, and meaningful activity.

The psychological impact of abundance is also influenced by societal support systems. Access to mental health care, educational resources, mentorship, and technology is crucial for navigating the transition successfully. Humans must advocate for inclusive structures that empower all members of society to thrive in a post-labor context. Flourishing emerges when societal investment ensures opportunity, guidance, and emotional well-being for everyone.

Existential reflection becomes more prominent when work is no longer required for survival. Humans have the space to ask deeper questions about meaning, purpose, and legacy. Engaging in philosophy, arts, community service, and self-development can provide answers and guidance. Flourishing arises when humans embrace inquiry, reflection, and ethical engagement as part of daily life.

Concluding the discussion on the psychological impact of jobless abundance, it is clear that freedom from work is both liberating and challenging. Humans are presented with the opportunity to explore creativity, relationships, learning, and purpose in ways previously constrained by economic necessity. Flourishing arises when individuals cultivate resilience, intentionality, and self-awareness, allowing freedom to become a platform for growth rather than a source of anxiety.

Identity and meaning must be actively constructed. Without employment as the primary anchor, humans must define themselves through creativity, contribution, and personal development. Flourishing occurs when individuals recognize their intrinsic value and embrace diverse ways of contributing to society, cultivating purpose beyond traditional labor.

Society plays a crucial role in supporting mental well-being. Access to mental health resources, mentorship, and community networks is vital to help individuals navigate the uncertainty and emotional challenges of jobless abundance. Humans must prioritize collective support systems that foster psychological resilience and social cohesion. Flourishing emerges when well-being is integrated into societal design, not treated as secondary to productivity.

AI offers unprecedented opportunities to enhance human life, but humans must use it wisely. Machines can free time, support learning, and facilitate collaboration, yet they cannot replace human judgment, creativity, or emotional connection. Flourishing arises when technology is used intentionally to augment human potential while maintaining autonomy and purpose.

Finally, the transition to jobless abundance invites reflection on the essence of human life. Freed from the constraints of survival-driven labor, humans have the chance to explore curiosity, ethics, relationships, and self-actualization. Flourishing emerges when individuals and societies cultivate environments that nurture meaning, connection, creativity, and mental well-being, ensuring that the psychological benefits of abundance are fully realized.

Inspirational Quote:

"Freedom from work is not freedom from purpose; it is an invitation to redefine what truly matters in life." — Anonymous

Part VI — Skills for the Next Decade

Chapter . 25

Learning How to Learn Again

Adaptability becomes the most valuable skill of all.

In a rapidly changing, AI-driven world, the ability to learn continuously is no longer optional—it is essential. Traditional education often emphasizes memorization and mastery of static knowledge, but the future demands adaptability, critical thinking, and the capacity to acquire new skills quickly. Humans must unlearn outdated assumptions about learning and embrace methods that foster curiosity, reflection, and experimentation. Flourishing arises when individuals develop lifelong learning habits that allow them to navigate complexity with confidence and creativity.

Learning how to learn again begins with mindset. Humans must cultivate intellectual humility, curiosity, and resilience, recognizing that knowledge is dynamic and constantly evolving. Accepting failure as part of the learning process allows individuals to explore new ideas without fear. Flourishing occurs when people view challenges as opportunities to grow rather than threats to competence or identity.

The role of AI in education and learning is transformative. Intelligent tutoring systems, personalized learning platforms, and adaptive algorithms can provide tailored feedback, simulate complex scenarios, and accelerate skill acquisition. Humans must use AI as a tool to enhance understanding and insight while maintaining critical judgment and creativity. Flourishing emerges when technology amplifies human learning rather than replacing the effort, curiosity, and reflection that make learning meaningful.

Collaboration enhances the process of learning again. Humans learn not only through individual effort but also through dialogue, mentorship, and shared problem-solving. AI can facilitate collaboration by connecting learners, suggesting resources, and tracking progress, creating rich networks of knowledge exchange. Flourishing arises when learning is both individual and collective, empowering humans to share insights and grow together.

Relearning how to learn also means embracing interdisciplinary thinking. AI and automation blur the boundaries between fields, making it essential to integrate knowledge from multiple domains. Humans must cultivate curiosity that spans science, art, philosophy, and technology, connecting insights to solve complex, real-world problems. Flourishing arises when individuals can think across disciplines, seeing patterns and opportunities that specialized knowledge alone cannot reveal.

Reflection and metacognition are critical components of learning again. Humans must develop the ability to assess their own thinking, recognize cognitive biases, and evaluate the effectiveness of different learning strategies. This self-awareness enables continuous improvement and adaptation. Flourishing occurs when individuals not only acquire knowledge but also understand how and why they learn, fostering autonomy and deeper insight.

Failure becomes a powerful teacher in this new paradigm. In a world where AI can handle routine tasks and experimentation is safer, humans can take calculated risks without fear of irreversible consequences. Mistakes become opportunities to iterate, explore, and innovate. Flourishing emerges when individuals embrace failure as a natural part of learning, cultivating resilience, creativity, and courage.

The social dimension of learning is equally important. Mentorship, collaboration, and community engagement provide context, feedback, and accountability. AI can facilitate these connections by matching learners with peers and experts, simulating real-world challenges, and tracking progress collaboratively. Flourishing arises when learning is enriched by shared experience, dialogue, and social support, combining human and technological resources for maximum growth.

Learning how to learn again also requires unlearning outdated habits and assumptions. Humans often rely on rote memorization, linear thinking, and passive absorption of information, which are insufficient in a world dominated by AI and rapid change. Embracing active learning, experimentation, and critical questioning is essential for adapting to evolving knowledge landscapes. Flourishing arises when individuals break free from rigid thinking and cultivate curiosity, adaptability, and resilience.

Technology offers both opportunity and challenge in this process. AI-driven tools can provide personalized guidance, simulate complex environments, and accelerate skill acquisition. However, humans must avoid overreliance on machines, ensuring that they retain critical thinking, creativity, and judgment. Flourishing occurs when humans integrate technology thoughtfully, using it to expand capability rather than outsource intelligence entirely.

Learning again involves cultivating patience and persistence. Mastery is rarely instantaneous, and deep understanding requires sustained effort and reflection. Humans must embrace iterative learning cycles, acknowledging that growth emerges from repeated engagement, feedback, and adjustment. Flourishing arises when individuals persist through challenges, balancing ambition with reflection and resilience.

The emotional dimension of learning is often overlooked. Motivation, curiosity, and self-confidence influence how effectively humans absorb, retain, and apply knowledge. Developing strategies to manage stress, maintain focus, and sustain engagement is critical in a rapidly changing environment.

Flourishing occurs when learning incorporates emotional intelligence, ensuring that the mind and heart are aligned in pursuit of growth.

Learning how to learn again emphasizes adaptability as the core skill for the 21st century. The rapid pace of technological and social change makes it impossible to rely solely on static knowledge or fixed skill sets. Humans must cultivate the ability to pivot, synthesize new information, and apply lessons in novel contexts. Flourishing arises when individuals embrace flexibility, experimentation, and continuous growth as fundamental to their development.

Mentorship and collaboration accelerate the learning process. Humans can benefit from sharing knowledge, receiving feedback, and observing diverse approaches to problem-solving. AI can facilitate these interactions by connecting learners, curating resources, and simulating complex scenarios. Flourishing occurs when humans combine personal initiative with social learning, creating networks of insight and support that amplify growth.

Self-directed learning becomes increasingly vital. In a world of abundant information and AI tools, humans must identify priorities, set goals, and maintain motivation. Learning is no longer confined to classrooms or workplaces; it occurs across experiences, interactions, and experimentation. Flourishing arises when individuals take ownership of their learning journey, balancing curiosity, reflection, and disciplined practice.

Ethical reasoning and critical thinking are central to learning again. With AI shaping knowledge, analysis, and decision-making, humans must evaluate sources, detect bias, and question assumptions. Learning is incomplete without understanding the implications and responsibilities associated with knowledge. Flourishing occurs when humans cultivate discernment and ethical awareness alongside technical and cognitive skills.

Concluding the discussion on learning how to learn again, it is clear that adaptability and curiosity are the most valuable skills for an AI-driven world. Traditional education and skill acquisition are insufficient; humans must embrace continuous, self-directed, and reflective learning to navigate uncertainty. Flourishing arises when individuals develop the capacity to grow, pivot, and engage meaningfully with an ever-changing landscape.

The integration of AI into learning offers immense potential, but humans must maintain agency and critical thinking. Machines can provide guidance, simulate experiences, and accelerate skill acquisition, yet the learner's curiosity, judgment, and creativity remain essential. Flourishing occurs when humans use technology to enhance understanding while preserving autonomy and insight.

Social and collaborative dimensions enrich the process. Learning in isolation limits perspective and adaptability, whereas interaction with mentors, peers, and communities cultivates deeper understanding and innovation. AI can facilitate these connections, creating opportunities for shared knowledge and cross-disciplinary insight. Flourishing emerges when humans combine personal initiative with collective wisdom, learning from and with others.

Persistence, reflection, and emotional engagement are key. True learning requires effort, patience, and resilience in the face of failure or uncertainty. Humans must nurture emotional intelligence alongside cognitive skills to sustain motivation, focus, and adaptability. Flourishing occurs when individuals embrace the full spectrum of learning, integrating mind, heart, and experience.

Finally, learning how to learn again is inseparable from purpose. Knowledge gains value when applied to meaningful goals, creativity, ethical action, and societal contribution. Humans must align learning with intention, curiosity, and the pursuit of impact. Flourishing arises when the journey of learning empowers individuals to navigate complexity, create meaning, and contribute to a world in constant transformation.

Inspirational Quote:

"*In a world that never stops changing, the most valuable skill is the courage and curiosity to keep learning*." — Anonymous

Chapter . 26

Critical Thinking in a World of Instant Answers

Discernment matters more when information is effortless.

In the age of AI, information is abundant, instantaneous, and often overwhelming. Humans no longer need to search for answers—they are delivered on demand, curated by algorithms and automated systems. While this creates convenience and efficiency, it also introduces risks: misinformation, cognitive overload, and superficial understanding. Critical thinking is essential to navigate this landscape, enabling humans to evaluate sources, assess validity, and apply knowledge wisely. Flourishing arises when discernment becomes a guiding principle, ensuring that access to information leads to understanding rather than confusion.

The challenge of instant answers lies in the human tendency to accept information at face value. Algorithms optimize for engagement, speed, or relevance, not truth or depth. Humans must cultivate skepticism, analytical reasoning, and the ability to question assumptions. Flourishing occurs when individuals combine curiosity with careful evaluation, distinguishing between data that informs and noise that misleads.

Critical thinking also involves recognizing bias—in AI, in human sources, and in one's own perception. Algorithms reflect the values and prejudices of their creators, while humans carry cognitive biases that shape interpretation. By developing metacognitive awareness and reflective habits, individuals can identify distortions and make more informed judgments. Flourishing emerges when humans engage critically, remaining aware of the limitations and influences shaping the information they consume.

Collaboration and dialogue enhance critical thinking. Engaging with diverse perspectives challenges assumptions, deepens understanding, and reveals hidden biases. AI can facilitate this process by connecting individuals with complementary knowledge, providing contrasting viewpoints, and synthesizing complex data. Flourishing occurs when humans integrate social learning with personal reflection, creating a richer, more nuanced understanding of the world.

The abundance of information can lead to cognitive overload, where humans struggle to filter relevance, prioritize learning, and retain insight. Critical thinking equips individuals with strategies to organize, evaluate, and synthesize data effectively. Humans must develop habits such as questioning assumptions, cross-referencing sources, and reflecting on conclusions. Flourishing occurs when individuals manage the flow of information intentionally, transforming abundance into understanding rather than overwhelm.

Decision-making in a world of instant answers requires balancing speed with reflection. AI can provide rapid solutions, but humans must consider context, values, and long-term consequences. Critical thinking ensures that decisions are not dictated solely by algorithmic outputs but informed by ethical reasoning and situational awareness. Flourishing arises when humans integrate intelligence, judgment, and foresight into everyday choices, preserving agency amidst technological efficiency.

Education and skill development must adapt to emphasize discernment. Memorizing facts is no longer sufficient; students must learn how to question, evaluate, and synthesize information critically. AI can enhance this learning by presenting complex scenarios, simulating problem-solving, and providing immediate feedback. Humans must embrace these tools while retaining independent judgment. Flourishing occurs when education cultivates thinkers capable of navigating complexity, uncertainty, and rapid change.

Social responsibility is closely tied to critical thinking. Humans influence each other through communication, media, and technology. Evaluating sources, questioning narratives, and assessing intentions become moral as well as cognitive tasks. Flourishing arises when individuals combine critical evaluation with ethical reflection, ensuring that their decisions and actions positively contribute to society and avoid harm.

Critical thinking in a world of instant answers also requires understanding context. Facts alone are insufficient; humans must interpret information within historical, cultural, and social frameworks. AI may provide data points and patterns, but humans are responsible for connecting the dots, recognizing nuances, and understanding implications. Flourishing arises when individuals integrate knowledge with context, ensuring that insight leads to meaningful understanding and action.

Analytical reasoning is complemented by creativity in problem-solving. AI can offer solutions based on patterns and logic, yet novel challenges often require imagination, intuition, and lateral thinking. Humans must combine critical analysis with creative approaches to address unprecedented problems. Flourishing occurs when humans leverage both rationality and imagination, enhancing decision-making in complex, dynamic environments.

Bias recognition extends to personal habits and societal systems. Critical thinkers must examine how their own assumptions, cultural conditioning, and social influences shape interpretation of information. AI itself may perpetuate systemic bias, requiring humans to evaluate not just the data but the structures behind it. Flourishing emerges when individuals cultivate self-awareness, ethical sensitivity, and vigilance in assessing both information and the frameworks producing it.

The iterative nature of critical thinking is essential. Humans must approach problems as ongoing inquiries, testing hypotheses, seeking feedback, and revising conclusions. AI can assist by modeling scenarios, analyzing outcomes, and suggesting alternative strategies, but humans must interpret, judge, and apply these insights responsibly. Flourishing occurs when learning and thinking are dynamic, reflexive processes rather than static or superficial exercises.

Critical thinking also entails evaluating the credibility and intent behind information sources. In a world of AI-generated content, deepfakes, and algorithmically curated news, humans must question who created the information, why it exists, and how it might be manipulated. Understanding context, purpose, and bias is essential for responsible decision-making. Flourishing arises when individuals combine analytical rigor with ethical awareness, safeguarding both personal and societal well-being.

Ethics and discernment are inseparable in the age of instant answers. Humans must not only assess truth but also consider consequences, fairness, and broader social impact. AI can provide rapid analyses, but the ethical interpretation remains a uniquely human responsibility. Flourishing occurs when humans apply critical thinking to align knowledge and action with moral principles, ensuring that technology serves human flourishing rather than undermining it.

Teaching and learning critical thinking must be a lifelong endeavor. With information evolving faster than any individual can master, humans must develop habits that foster curiosity, reflection, and adaptability. AI tools can assist by presenting complex scenarios, analyzing patterns, and offering feedback, but the cultivation of judgment and discernment remains fundamentally human. Flourishing arises when continuous learning is integrated with reflective and ethical thinking.

Collaboration amplifies critical thinking by exposing individuals to diverse perspectives, feedback, and debate. AI can support these collaborative efforts by connecting experts, analyzing data, and facilitating communication across distances. Humans, however, must integrate this input thoughtfully, weighing evidence, context, and potential biases. Flourishing occurs when individuals balance collective insight with personal judgment, creating well-rounded understanding.

Concluding the discussion on critical thinking in a world of instant answers, it is clear that access to information alone does not guarantee understanding or wisdom. Humans must cultivate discernment, context awareness, ethical reasoning, and reflective judgment to navigate the flood of data presented by AI and digital systems. Flourishing arises when individuals balance curiosity with skepticism, analysis with creativity, and speed with careful consideration.

Critical thinking requires ongoing practice and self-awareness. Humans must recognize cognitive biases, question assumptions, and engage with diverse perspectives to reach robust conclusions. AI can support this process by providing data, simulations, and insights, but human judgment remains indispensable. Flourishing occurs when individuals integrate analytical rigor with ethical awareness and creative problem-solving, applying knowledge responsibly.

Collaboration enhances critical thinking. Engaging in dialogue, debate, and cooperative problem-solving exposes individuals to alternative viewpoints and strengthens understanding. AI can facilitate these interactions by connecting learners, aggregating information, and suggesting novel approaches. Flourishing arises when humans combine personal reflection with collective intelligence, generating solutions that are both informed and ethically grounded.

Resilience and humility are essential companions to critical thinking. Humans must accept uncertainty, ambiguity, and evolving knowledge while maintaining curiosity and ethical responsibility. Flourishing occurs when individuals embrace challenges, refine understanding through iterative inquiry, and remain open to new evidence and perspectives.

Finally, critical thinking in the age of instant answers is inseparable from purpose. Knowledge gains value when applied thoughtfully, ethically, and creatively to real-world challenges. Humans must align insight with action, ensuring that understanding translates into responsible decision-making and meaningful contribution. Flourishing emerges when discernment, ethics, and action converge, empowering humans to navigate complexity with wisdom, integrity, and impact.

Author Reflection:

Writing this chapter highlighted that in a world overflowing with information, knowledge alone is insufficient. True flourishing depends on the ability to think critically, question deeply, and act ethically. Critical thinking becomes the compass guiding humans through a landscape of instant answers and algorithmic influence.

Inspirational Quote:

"*Information is abundant; wisdom is rare. Cultivate discernment, and you will thrive in a world of endless answers.*" — Anonymous

Chapter . 27

Ethical Reasoning as a Practical Skill

Moral judgment becomes essential in AI-mediated decisions.

As AI increasingly shapes decisions in business, healthcare, education, and governance, humans must cultivate ethical reasoning as a core skill. Machines can process data, optimize outcomes, and predict patterns, but they cannot determine right from wrong, justice from exploitation, or long-term societal benefit from short-term gain. Humans must actively evaluate the consequences, values, and fairness of AI-driven choices. Flourishing arises when moral judgment guides both personal and societal decisions, ensuring that technology enhances human well-being rather than undermining it.

Ethical reasoning begins with awareness of consequences. Decisions made by AI can have far-reaching effects on individuals, communities, and global systems. Humans must anticipate potential harms, unintended outcomes, and distributional effects, considering who benefits and who may be disadvantaged. Flourishing occurs when humans integrate foresight and empathy into decision-making, promoting equitable and responsible outcomes.

Values play a central role in ethical reasoning. Humans must identify core principles—such as justice, dignity, autonomy, and sustainability—and use them as benchmarks for evaluating choices. AI systems can be programmed to reflect these principles, but humans remain responsible for defining and enforcing them. Flourishing arises when decisions align with deeply held ethical standards, balancing efficiency with humanity.

Ethical reasoning also requires critical reflection on biases and assumptions. AI systems often reflect the values and prejudices of their creators, making it essential for humans to scrutinize the underlying frameworks. By combining critical thinking with moral reflection, humans can identify and correct for systemic bias, ensuring fairness and inclusion. Flourishing emerges when individuals and organizations actively confront bias and prioritize equity in decision-making.

Ethical reasoning in an AI-driven world requires balancing competing interests. Decisions often involve trade-offs between efficiency, equity, privacy, and innovation. Humans must weigh short-term gains against long-term consequences, considering the needs of diverse stakeholders. Flourishing arises when individuals and organizations make choices that prioritize fairness, sustainability, and the common good, even when doing so is complex or inconvenient.

Transparency is essential for ethical practice. AI systems can operate as opaque "black boxes," making it difficult to understand how decisions are made. Humans must demand clarity, auditability, and explainability to ensure accountability. Flourishing occurs when people actively engage with technology, questioning and verifying decisions rather than accepting outcomes passively.

Collaboration enhances ethical reasoning. Complex moral challenges require input from multiple perspectives, including diverse cultural, professional, and philosophical viewpoints. AI can support this process by aggregating insights, modeling scenarios, and highlighting potential impacts. Humans, however, remain responsible for synthesizing information, deliberating values, and making judgments. Flourishing arises when collective intelligence and ethical reflection are combined to guide action responsibly.

Ethical reasoning also depends on emotional awareness. Empathy, compassion, and understanding of human experiences are crucial for evaluating the impact of decisions. Machines can analyze patterns and predict outcomes, but they cannot feel the human consequences of their actions. Flourishing occurs when humans integrate emotional intelligence with analytical reasoning, ensuring that moral judgment considers both logic and lived experience.

Ethical reasoning requires a deep understanding of power and responsibility. AI systems often concentrate influence in the hands of corporations, governments, or a few individuals, shaping outcomes that affect millions. Humans must question who holds decision-making authority and ensure accountability mechanisms are in place. Flourishing arises when ethical reflection informs both personal choices and societal structures, preventing abuse of power and promoting fairness.

Decision-making frameworks can help operationalize ethical reasoning. Approaches such as utilitarian analysis, rights-based evaluation, virtue ethics, and care ethics provide structured methods to assess consequences, obligations, and values. Humans must learn to apply these frameworks flexibly, recognizing that no single approach suffices for all situations. Flourishing occurs when individuals navigate moral complexity with clarity, rigor, and compassion.

Ethical reasoning is closely tied to trust. Individuals, organizations, and societies must trust that AI systems are aligned with shared values and that decisions reflect ethical considerations. Humans must engage in oversight, evaluation, and dialogue to maintain this trust. Flourishing arises when ethical practices foster confidence in AI and human decision-making, creating systems that support collective well-being.

Global perspectives are increasingly important. AI-driven decisions often have transnational effects, influencing economies, cultures, and environments beyond local contexts. Humans must consider diverse cultural norms, international laws, and global consequences when evaluating ethical dilemmas. Flourishing occurs when ethical reasoning incorporates both local understanding and global awareness, promoting justice and responsibility across borders.

Ethical reasoning also demands vigilance against unintended consequences. AI systems can amplify harm even when designed with good intentions, from biased hiring algorithms to predictive policing that reinforces inequality. Humans must anticipate potential downstream effects, monitor outcomes, and adjust practices as needed. Flourishing arises when foresight and accountability are embedded into every stage of AI development and deployment.

Transparency and explainability are crucial in operationalizing ethical reasoning. Humans need to understand how AI systems make decisions, why certain recommendations are produced, and what assumptions underlie algorithmic outputs. By interrogating these processes, individuals can ensure alignment with ethical principles. Flourishing occurs when humans actively seek clarity, demand accountability, and maintain oversight over automated systems.

Ethical reasoning also intersects with creativity. Designing AI solutions that are both effective and morally sound requires imaginative approaches to complex problems. Humans must envision alternative scenarios, explore innovative safeguards, and consider novel ways to balance competing interests. Flourishing emerges when ethical reflection fuels creativity, generating solutions that are both responsible and transformative.

Education plays a pivotal role in cultivating practical ethical reasoning. Teaching students, professionals, and leaders to integrate moral reflection with problem-solving equips society to navigate AI's challenges. Ethical training, real-world case studies, and interdisciplinary approaches develop the skills necessary for informed, responsible decision-making. Flourishing occurs when ethical competence becomes a standard, expected, and celebrated aspect of human development.

Concluding the discussion on ethical reasoning as a practical skill, it is clear that moral judgment is essential in a world increasingly influenced by AI. Machines can optimize, predict, and analyze, but they cannot make value-based decisions or consider the ethical implications of their outputs. Humans must actively engage with AI, guiding its deployment with discernment, empathy, and a commitment to fairness. Flourishing arises when ethical reasoning informs both personal choices and societal governance, ensuring technology serves humanity rather than undermining it.

Practical ethical reasoning requires continuous reflection and vigilance. Humans must evaluate decisions, recognize biases, anticipate consequences, and adapt frameworks as challenges evolve. Ethical principles should guide design, policy, and daily actions, integrating foresight, justice, and compassion. Flourishing occurs when moral reflection translates into responsible behavior and impactful choices, creating systems that enhance collective well-being.

Collaboration and dialogue strengthen ethical practice. Engaging with diverse perspectives, cultural contexts, and interdisciplinary knowledge improves understanding and decision-making. AI can support this process by modeling scenarios, highlighting trade-offs, and connecting stakeholders. Humans, however, must synthesize insights, deliberate values, and act with integrity. Flourishing arises when collaborative ethical reasoning shapes AI and human decision-making in harmony.

Courage and accountability are critical components of ethical reasoning. Upholding moral principles in complex, high-pressure situations requires resilience and commitment. Humans must be willing to challenge unethical practices, intervene in harmful processes, and advocate for fairness and justice. Flourishing occurs when ethical courage becomes the norm, guiding both individual and collective action in a rapidly evolving world.

Finally, ethical reasoning is inseparable from purpose. Decisions guided by values, foresight, and empathy allow humans to align knowledge, action, and technology with meaningful impact. Flourishing emerges when humans integrate ethical reflection into daily life, fostering systems, behaviors, and AI that enhance dignity, equity, and human potential.

Inspirational Quote:

"*Ethics is not a luxury of thought—it is the compass that guides our actions in a world of limitless power.*" — Anonymous

Chapter . 28

Cultural Intelligence and Global Awareness

As AI and digital technologies connect people across borders, cultural intelligence has become an essential skill for human flourishing. Global interactions are faster, more frequent, and more complex than ever before. Yet while technology enables connection, it also magnifies misunderstandings, biases, and misinterpretations. Humans must cultivate the ability to navigate diverse cultural contexts, recognize differences, and communicate effectively across boundaries. Flourishing arises when individuals combine curiosity, empathy, and adaptability to engage respectfully and insightfully with the world.

Cultural intelligence begins with awareness of one's own assumptions and biases. Humans often interpret behavior, communication, and values through the lens of their own culture, risking misjudgment or offense. Self-reflection, education, and openness to feedback help individuals recognize these blind spots. Flourishing occurs when self-awareness guides interactions, allowing humans to bridge gaps and foster mutual understanding.

Empathy is central to cultural intelligence. Understanding others' perspectives requires more than observation; it demands listening, compassion, and the willingness to suspend judgment. Humans must engage with humility, acknowledging that their own worldview is not universal. Flourishing arises when empathy guides cross-cultural relationships, promoting cooperation, respect, and shared problem-solving.

Global awareness also entails knowledge of historical, social, and political contexts. Actions that seem innocuous in one context can have unintended consequences elsewhere. AI can provide access to data, simulations, and patterns to inform understanding, but humans must interpret this knowledge thoughtfully. Flourishing occurs when awareness of context informs ethical, responsible, and culturally sensitive decisions.

Cultural intelligence also involves navigating the ethical dimensions of global interaction. What is acceptable in one society may be considered unethical or harmful in another. Humans must balance respect for cultural norms with universal principles of fairness, dignity, and human rights. AI can assist by providing data and insights, but ethical interpretation remains a human responsibility. Flourishing arises when humans harmonize cultural sensitivity with moral judgment, fostering cooperation without compromising values.

Language is both a bridge and a barrier. Miscommunication can result from literal translation, idiomatic differences, or nuanced social cues. Humans must develop skills in listening, observation, and contextual interpretation to communicate effectively across linguistic and cultural divides. AI can facilitate translation and pattern recognition, but human judgment is needed to capture tone, intent, and subtle meaning. Flourishing occurs when humans leverage both technology and empathy to create meaningful cross-cultural dialogue.

Cross-cultural teamwork is another key component of global awareness. Projects involving multinational teams require collaboration across time zones, norms, and expectations. AI can coordinate schedules, track progress, and suggest optimal workflows, but humans must navigate interpersonal dynamics, resolve conflicts, and build trust. Flourishing arises when humans cultivate collaboration skills, respect differences, and use AI to enhance, rather than replace, human connection.

Global awareness also involves recognizing power dynamics and historical context. Inequalities, colonial legacies, and systemic discrimination influence interactions across borders. Humans must approach collaboration with humility and a commitment to equity, ensuring that engagement does not perpetuate harm. Flourishing occurs when individuals integrate awareness of historical and social realities into cross-cultural relationships, promoting justice alongside cooperation.

Cultural intelligence also requires humility and curiosity. Humans must accept that they do not have all the answers and that each culture offers unique insights, values, and approaches to life's challenges. Approaching cross-cultural interactions with a learner's mindset allows for richer understanding and collaboration. Flourishing arises when humility and curiosity guide human engagement, transforming differences into opportunities for growth.

AI-mediated communication presents both opportunities and challenges for cultural intelligence. While AI can translate languages, detect sentiment, and model social norms, it can also amplify biases embedded in datasets. Humans must critically evaluate AI outputs and ensure they align with ethical, culturally sensitive practices. Flourishing occurs when humans use AI responsibly, combining technological tools with judgment, empathy, and contextual awareness.

Building trust across cultures is essential for effective collaboration. Trust emerges through consistent behavior, clear communication, and respect for local norms and practices. AI can support trust by providing transparency, verifying facts, and facilitating accountability, but human engagement remains the core of relational bonds. Flourishing arises when humans cultivate trust intentionally, creating relationships that withstand complexity and uncertainty.

Conflict resolution is a key aspect of global awareness. Misunderstandings, differing values, and competing interests can create tension in multicultural settings. Humans must develop skills in negotiation, mediation, and empathy to navigate conflicts constructively. AI can provide simulations or data analysis to inform resolution strategies, but humans must lead with emotional intelligence and ethical reasoning. Flourishing occurs when differences are approached as opportunities for dialogue and problem-solving, not sources of division.

Cultural intelligence also requires emotional awareness and regulation. In cross-cultural interactions, emotions can be misinterpreted due to differing norms and expressions. Humans must develop sensitivity to nonverbal cues, tone, and context while managing their own emotional responses. AI tools can provide feedback on communication patterns, but emotional judgment remains a uniquely human skill. Flourishing occurs when emotional intelligence complements cultural understanding, fostering harmony and mutual respect.

Ethical dilemmas often emerge in global interactions. Practices considered acceptable in one culture may conflict with universal human rights or ethical standards. Humans must navigate these tensions thoughtfully, balancing respect for diversity with moral responsibility. AI can assist by modeling scenarios or analyzing impacts, yet humans must make principled decisions. Flourishing arises when ethical reasoning and cultural intelligence work together to guide responsible action.

Continuous reflection is vital for sustaining cultural awareness. Humans must periodically examine their assumptions, biases, and habits to prevent misunderstandings and promote inclusivity. Engaging in dialogue, receiving feedback, and learning from mistakes strengthen cross-cultural competence. Flourishing occurs when reflection and adaptation become habits, allowing humans to grow through ongoing interaction and learning.

Cultural intelligence is also enhanced by immersion and experience. Travel, study abroad, global collaboration, and diverse social networks provide rich opportunities to deepen understanding. Humans must seek experiences that challenge assumptions and expand perspectives. AI can curate learning opportunities and facilitate access to global knowledge, but lived experience remains irreplaceable. Flourishing arises when humans actively engage with diverse cultures, integrating lessons into thought and action.

Concluding the discussion on cultural intelligence and global awareness, it is clear that flourishing in an interconnected world requires more than knowledge—it demands empathy, reflection, and adaptability. Humans must navigate diverse perspectives, norms, and ethical frameworks while using AI to support understanding rather than replace human judgment. Flourishing arises when cultural intelligence becomes a habitual practice, guiding ethical and creative engagement across borders.

Global awareness also strengthens collaboration and innovation. Exposure to different viewpoints, problem-solving approaches, and cultural practices enhances creativity and enables humans to tackle complex challenges more effectively. AI can facilitate this by providing data, insights, and translation, but humans must integrate these tools with ethical judgment and interpersonal skills. Flourishing occurs when humans combine technological support with emotional and cultural intelligence to co-create solutions for shared challenges.

Cultural intelligence nurtures resilience in the face of uncertainty and change. Humans encounter unfamiliar norms, language barriers, and conflicting values, and success requires flexibility, patience, and adaptability. Flourishing arises when humans approach difference as an opportunity to learn and grow, transforming potential obstacles into sources of insight and connection.

Continuous learning, reflection, and feedback are essential for maintaining global competence. Humans must actively engage in practices that challenge assumptions, expand understanding, and refine intercultural skills. AI can support learning through simulations, scenario planning, and curated content, but humans must apply critical thinking and empathy to synthesize these lessons. Flourishing occurs when learning and action reinforce each other, allowing individuals to navigate complexity with skill and integrity.

Finally, cultural intelligence is inseparable from purpose and ethical engagement. Humans must use their understanding of diverse cultures to foster equity, justice, and collaboration while leveraging AI responsibly. Flourishing emerges when humans bridge divides, create inclusive spaces, and apply insight to meaningful action, ensuring that global interconnectedness strengthens both individuals and societies.

Author Reflection:

Writing this chapter underscored that global awareness is not merely knowledge—it is a practice of humility, empathy, and continuous learning. AI can connect us, but humans must cultivate judgment, emotional intelligence, and ethical awareness to navigate cultural complexity effectively. True flourishing occurs when these skills are integrated into daily life and global collaboration.

Inspirational Quote:

"*Understanding the world requires listening beyond our own perspective; empathy and curiosity are the keys to global connection.*" — Anonymous

Chapter . 29

Creativity, Curiosity, and Courage

Human originality thrives where machines cannot feel or wonder.

In a world where AI can generate art, compose music, and solve complex problems, human creativity becomes an irreplaceable asset. Machines can mimic patterns and optimize processes, but they lack the lived experience, intuition, and emotional depth that inspire truly original ideas. Humans must cultivate creativity deliberately, combining curiosity, imagination, and courage to explore the unknown. Flourishing arises when individuals embrace the power of human originality, using it to create, innovate, and contribute meaningfully to society.

Curiosity is the spark that drives creativity. Humans must remain inquisitive, questioning assumptions, exploring new domains, and seeking connections across disciplines. AI can provide vast resources and insights to fuel exploration, but the desire to discover and experiment must come from the human mind. Flourishing occurs when curiosity leads to continual learning, experimentation, and the pursuit of ideas that expand understanding and possibility.

Courage is essential in translating creativity into action. Original ideas often involve risk, uncertainty, and potential failure. Humans must overcome fear, self-doubt, and external skepticism to bring innovative concepts to life. AI can assist by providing simulations, analysis, and optimization, but courage is inherently human. Flourishing arises when individuals act boldly, embracing risk as part of the creative process and learning from setbacks along the way.

Collaboration enhances creativity by combining diverse perspectives and experiences. Humans working together can generate ideas that surpass what any individual or machine could achieve alone. AI can facilitate collaboration by connecting people across geographies, providing insights, and suggesting novel solutions. Flourishing occurs when collaboration, guided by curiosity and courage, produces innovations that benefit both individuals and communities.

Creativity requires the freedom to experiment and take intellectual risks. In environments where failure is stigmatized or discouraged, curiosity is stifled and innovation stalls. Humans must cultivate spaces—both personal and professional—where exploration is encouraged, mistakes are treated as learning opportunities, and unconventional ideas are valued. Flourishing occurs when curiosity is nurtured in a supportive culture that prioritizes growth over perfection.

Observation and reflection are key components of creative thinking. Humans must notice patterns, question assumptions, and reflect deeply on experiences to generate novel ideas. AI can support this process by analyzing trends, identifying anomalies, and suggesting possibilities, but reflection remains inherently human. Flourishing arises when humans balance analysis with intuition, combining machine insights with lived experience and imaginative thought.

Interdisciplinary thinking fuels originality. Creative breakthroughs often occur at the intersections of fields, cultures, or concepts. Humans must explore beyond familiar domains, seeking inspiration in art, science, philosophy, and everyday life. AI can provide tools for research and synthesis, but curiosity drives the connections that lead to transformative ideas. Flourishing occurs when humans integrate diverse perspectives, generating solutions that are both innovative and impactful.

Emotional engagement is another vital aspect of creativity. Human experiences, feelings, and values inform art, design, innovation, and problem-solving in ways machines cannot replicate. AI may mimic style or structure, but it lacks consciousness, empathy, and passion. Flourishing arises when humans draw upon emotion and lived experience to create works and solutions that resonate with meaning and purpose.

Creativity thrives when humans embrace ambiguity and uncertainty. Unlike AI, which seeks predictable patterns, humans excel in navigating the unknown and generating novel solutions in complex, uncertain situations. Embracing ambiguity requires flexibility, adaptability, and openness to possibilities that cannot be fully anticipated. Flourishing arises when humans see uncertainty not as a threat, but as fertile ground for innovation and discovery.

Curiosity drives the exploration of uncharted territories. Humans must ask questions that challenge conventional wisdom, probe hidden assumptions, and seek connections others might overlook. AI can provide data, patterns, and suggestions, but it cannot initiate genuine wonder or a desire to explore the unknown. Flourishing occurs when curiosity motivates continuous learning and imaginative problem-solving, expanding human potential in ways machines alone cannot achieve.

Courage enables humans to transform creative ideas into action. Original concepts often face resistance, skepticism, or institutional barriers. Acting on one's vision requires confidence, persistence, and a willingness to confront uncertainty. AI can support decision-making and simulate outcomes, but courage must come from the human heart and mind. Flourishing arises when individuals combine bravery with imagination, bringing bold ideas into reality.

Collaboration amplifies creativity. When humans work together, they pool diverse experiences, perspectives, and skills to generate solutions that surpass individual efforts. AI can enhance this process by facilitating communication, connecting collaborators, and modeling scenarios, but human insight and judgment remain central. Flourishing occurs when creativity is nurtured collectively, producing innovations that are ethically sound, socially relevant, and emotionally resonant.

Creativity, curiosity, and courage are inseparable from ethical responsibility. Humans must ensure that their innovative ideas serve the greater good, rather than exploiting or harming others. AI can assist by analyzing potential consequences, highlighting risks, and modeling impact, but the ethical evaluation must come from humans. Flourishing arises when creativity is guided by moral reflection, combining ingenuity with integrity.

Diverse experiences fuel original thinking. Exposure to different cultures, disciplines, and viewpoints expands the range of ideas and possibilities. Humans must actively seek opportunities to learn from environments outside their comfort zones, leveraging curiosity to absorb lessons from the unfamiliar. AI can provide tools to access knowledge and perspectives worldwide, but humans must synthesize these insights to generate truly novel solutions. Flourishing occurs when diverse experiences inform creative and courageous action.

Failure is an essential companion to innovation. Many groundbreaking ideas emerge only after repeated attempts, experimentation, and adaptation. Humans must embrace failure as a learning tool, extracting insight and adjusting approaches accordingly. AI can simulate experiments and predict outcomes, but resilience—the ability to persevere through setbacks—remains a human strength. Flourishing arises when individuals combine courage, persistence, and reflection to transform failures into stepping stones for success.

Play and imagination are integral to human creativity. Unlike AI, humans can dream, imagine possibilities, and explore fantastical scenarios beyond the constraints of logic or optimization. Nurturing a sense of wonder, curiosity, and playfulness stimulates original thought and problem-solving. Flourishing occurs when individuals reclaim the freedom to experiment, imagine, and create, transforming both personal and societal life in the process.

Concluding the discussion on creativity, curiosity, and courage, it is clear that human originality remains indispensable in a world increasingly influenced by AI. Machines can generate patterns, analyze data, and simulate scenarios, but they cannot experience wonder, imagination, or ethical reflection. Humans must intentionally cultivate creativity, nurture curiosity, and summon courage to act on ideas, ensuring that technology complements rather than replaces human ingenuity. Flourishing arises when originality is guided by purpose, insight, and ethical responsibility.

The interplay of curiosity and courage fuels meaningful innovation. Curiosity drives exploration and discovery, while courage enables action despite uncertainty and risk. Together, they empower humans to push boundaries, challenge norms, and develop solutions that address complex, real-world problems. Flourishing occurs when these qualities are consistently nurtured, allowing human potential to thrive alongside AI capabilities.

Collaboration amplifies the impact of creative efforts. Engaging with diverse perspectives, cultures, and disciplines enriches ideas, enhances problem-solving, and strengthens outcomes. AI can support collaboration by providing insights, modeling scenarios, and facilitating connections, but human judgment, empathy, and communication remain central. Flourishing arises when creativity is both individual and collective, producing solutions that are socially, ethically, and culturally meaningful.

Resilience and reflection complete the cycle of creativity. Humans must view failure as a learning opportunity, integrate feedback, and iterate continuously. AI can assist in testing and optimizing ideas, but reflection ensures that innovation aligns with human values and purpose. Flourishing occurs when individuals embrace failure, learn from experience, and persist with courage and curiosity.

Finally, human creativity is inseparable from ethical and purposeful engagement. Flourishing arises when curiosity, courage, and imagination are channeled toward solving meaningful problems, inspiring others, and enhancing collective well-being. In a world where machines can replicate tasks and optimize systems, human originality remains the irreplaceable force that drives progress, culture, and human connection.

Author Reflection:

Writing this chapter reinforced that creativity is not optional—it is the essence of human flourishing. Curiosity inspires exploration, courage enables action, and imagination transforms ideas into meaningful contributions. AI may augment capabilities, but human originality, guided by ethics and purpose, is what truly shapes the future.

Inspirational Quote:

"Creativity is the human spark that machines cannot replicate; curiosity fuels it, courage sustains it, and purpose gives it meaning." — Anonymous

Part VII — The Next Generation

Chapter . 30

Children Growing Up with Thinking Machines

AI will shape how the next generation learns, plays, and imagines.

Children today are growing up surrounded by AI, from digital assistants and educational apps to immersive virtual worlds. Unlike previous generations, they interact with machines that can think, respond, and adapt to their behavior. This creates opportunities for personalized learning, creative exploration, and global connection, but it also raises questions about cognitive development, socialization, and identity formation. Flourishing arises when humans guide children to engage with AI thoughtfully, balancing technological benefits with human-centered growth.

The presence of AI in play and education reshapes attention and curiosity. Children can explore topics at their own pace, receive instant feedback, and experience simulations that were previously unimaginable. Yet, constant access to algorithmic guidance can also reduce independent problem-solving and imaginative exploration. Humans must help children develop agency, curiosity, and resilience, ensuring that AI complements rather than replaces critical thinking and creativity.

AI-mediated learning has the potential to democratize education. Children from diverse backgrounds can access high-quality resources, personalized instruction, and global knowledge networks. However, disparities in technology access and literacy can exacerbate existing inequalities. Flourishing occurs when families, educators, and policymakers ensure equitable access, teaching children not only to use AI but to understand and question it critically.

Social development is profoundly influenced by AI companions, chatbots, and virtual peers. Children may form attachments to digital entities, learn social norms through AI interactions, or rely on machines for emotional support. While these tools can enhance empathy and communication, humans must guide children in developing real-world relationships, emotional resilience, and ethical awareness. Flourishing arises when AI augments, rather than replaces, human connection and interpersonal growth.

The influence of AI on attention spans and learning habits is profound. Instant feedback, gamified experiences, and adaptive algorithms can accelerate skill acquisition, but they may also encourage superficial engagement or reliance on quick solutions. Children must be taught to balance AI-driven efficiency with deep focus, reflection, and persistence. Flourishing arises when humans guide children to use technology as a tool for growth rather than a crutch that diminishes resilience or creativity.

Ethical awareness is essential for children growing up with AI. From understanding data privacy to recognizing algorithmic bias, children must learn that AI systems are not neutral. They must develop the ability to question machine outputs, consider fairness, and understand the broader social implications of technology. Flourishing occurs when ethical reasoning is cultivated early, instilling values that guide responsible interaction with AI throughout life.

AI also impacts play, imagination, and creativity. Virtual worlds, intelligent toys, and interactive simulations offer unprecedented opportunities for exploration and storytelling. Yet, reliance on AI-guided play can limit open-ended creativity and problem-solving if children are not encouraged to experiment independently. Flourishing arises when humans balance structured AI experiences with unstructured, imaginative, and collaborative play that nurtures originality and curiosity.

Social skills require deliberate cultivation in an AI-rich environment. Children may form bonds with AI companions, potentially reducing motivation to engage in human relationships. Guidance from parents, educators, and communities is essential to ensure children develop empathy, teamwork, and conflict resolution. Flourishing occurs when AI supports social growth rather than replacing human interaction, helping children navigate both digital and real-world relationships.

Children growing up with AI are exposed to personalized learning environments that respond dynamically to their abilities and interests. This adaptability can accelerate understanding and provide motivation, but it may also create echo chambers that reinforce existing preferences and limit exposure to new ideas. Humans must guide children to engage with diverse content, perspectives, and challenges to foster broad thinking. Flourishing arises when AI is used to expand horizons rather than narrow them, nurturing curiosity, adaptability, and resilience.

The integration of AI into education also impacts the role of teachers. Educators are shifting from knowledge dispensers to facilitators of critical thinking, creativity, and ethical understanding. AI can handle repetitive instruction, track progress, and analyze learning patterns, freeing teachers to focus on mentoring, discussion, and exploration. Flourishing occurs when humans guide AI to augment educational experiences, ensuring that technology strengthens rather than replaces meaningful human guidance.

Digital literacy is essential for children in an AI-driven world. Understanding how algorithms function, recognizing bias, evaluating sources, and maintaining privacy are critical skills for navigating life safely and responsibly. Humans must equip children with the ability to critically engage with AI systems while fostering judgment and autonomy. Flourishing arises when digital competence is combined with ethical reasoning and reflective thinking, enabling children to interact with technology confidently and responsibly.

The impact of AI on play and socialization also extends to emotional development. Children may develop empathy for AI companions or rely on machines for problem-solving and emotional support. Parents and educators must ensure that children cultivate emotional intelligence, resilience, and the ability to engage meaningfully with peers. Flourishing occurs when AI enhances emotional growth rather than substituting for human relationships, allowing children to develop self-awareness, empathy, and social competence.

The integration of AI into children's daily lives also shapes their problem-solving skills. AI can suggest solutions, model outcomes, and provide hints, but children must learn to evaluate options, consider consequences, and exercise judgment. Overreliance on AI can hinder independent thinking, while guided use encourages strategic reasoning and creative approaches. Flourishing arises when children use AI as a tool to enhance their own problem-solving capacity rather than as a substitute for effort or reflection.

AI-mediated learning also influences collaboration and teamwork. Digital platforms connect children across classrooms, cities, and countries, creating opportunities for shared projects, communication, and collective creativity. Humans must guide children in negotiating differences, managing conflicts, and valuing diverse perspectives, ensuring that technology fosters cooperative skills. Flourishing occurs when AI supports meaningful collaboration while reinforcing empathy, communication, and mutual respect.

Curiosity-driven exploration remains central to human development. Children must be encouraged to ask questions, test ideas, and venture beyond AI-provided pathways. Machines can supply data, suggest experiments, and highlight patterns, but the desire to explore, imagine, and wonder must be cultivated consciously. Flourishing arises when curiosity motivates self-directed learning and inspires children to discover possibilities beyond algorithmic boundaries.

Ethical reflection becomes increasingly critical as AI mediates more aspects of childhood. Children must understand the implications of their interactions with AI, from privacy concerns to the fairness of algorithms and the potential impact of digital behavior on themselves and others. Humans must model ethical reasoning and encourage discussion, helping children develop moral frameworks to guide action. Flourishing occurs when ethical awareness becomes a habit, shaping responsible engagement with technology and society.

Concluding the discussion on children growing up with thinking machines, it is evident that AI presents both extraordinary opportunities and significant challenges for the next generation. Technology can enhance learning, creativity, and global connection, but it can also shape attention, social development, and ethical reasoning in ways that require careful guidance. Flourishing arises when adults actively support children in integrating AI responsibly, ensuring that machines amplify human potential rather than replace essential human experiences.

Parental and educational guidance is crucial. Adults must help children navigate AI-rich environments with critical thinking, curiosity, and ethical awareness. They must encourage exploration while modeling reflection, empathy, and resilience. Flourishing occurs when guidance fosters independent thought, creativity, and a sense of responsibility in children, empowering them to use AI thoughtfully throughout life.

The balance between AI use and human interaction is central to healthy development. Children need opportunities for unstructured play, real-world socialization, and imaginative exploration that extend beyond algorithmic suggestion or automated feedback. Flourishing arises when technology supports rather than dominates children's cognitive, emotional, and social growth, ensuring holistic development.

Ethical reasoning and digital literacy are foundational. Children must understand how AI works, its potential biases, and the implications of their digital actions. Adults can scaffold these lessons, modeling curiosity, discernment, and responsible engagement. Flourishing occurs when children develop the skills, judgment, and values necessary to navigate AI safely and meaningfully.

Finally, growing up with AI offers a unique chance to cultivate lifelong curiosity, creativity, and courage. By guiding children to engage critically, ethically, and imaginatively with technology, humans can empower the next generation to thrive in an increasingly complex and interconnected world. Flourishing arises when children embrace AI as a tool for exploration, learning, and connection while maintaining the uniquely human capacities that machines cannot replicate.

Author Reflection:

Writing this chapter highlighted that children are the first generation to experience AI as an integral part of daily life. Guiding them requires mindfulness, balance, and intentionality. True flourishing occurs when technology is harnessed to enhance human growth, imagination, and ethical awareness, preparing children to thrive in a future shaped by both humans and machines.

Inspirational Quote:

"*The goal is not to shield children from AI, but to teach them to walk wisely alongside it, guided by curiosity, courage, and conscience.*" — Anonymous

Chapter . 31

Education Beyond Memorization

Schools must prepare students for a world where answers are instant.

The traditional model of education—focused on memorization, repetition, and standardized testing—is increasingly outdated in a world where AI can provide instant answers. Students no longer need to store facts in their minds; instead, they must learn to understand, interpret, and apply information critically. Flourishing arises when education shifts from rote learning to developing skills such as analysis, creativity, problem-solving, and ethical reasoning, preparing students for a rapidly evolving world.

Critical thinking is central to this new model. Students must evaluate information, identify patterns, detect biases, and synthesize knowledge from multiple sources. AI can supply vast amounts of data, but humans must interpret it thoughtfully, considering context, reliability, and relevance. Flourishing occurs when learners are guided to apply understanding, rather than merely recall facts, ensuring meaningful engagement with knowledge.

Collaboration and communication skills are equally essential. Real-world problems rarely exist in isolation; students must work with others across disciplines, cultures, and contexts. AI can facilitate collaboration by connecting learners, providing insights, and organizing workflows, but humans must navigate interpersonal dynamics, negotiate solutions, and communicate ideas effectively. Flourishing arises when education fosters teamwork, empathy, and effective dialogue.

Creativity and innovation must be cultivated alongside critical thinking. Students need opportunities to explore, experiment, and generate original ideas. AI can support creative exploration by suggesting alternatives, modeling scenarios, or providing inspiration, but human imagination remains irreplaceable. Flourishing occurs when education emphasizes curiosity, experimentation, and the ability to generate new solutions, empowering learners to engage actively with complex challenges.

Education beyond memorization requires fostering adaptability. In a world where knowledge is constantly updated and AI generates new insights at unprecedented speed, students must learn to adjust, unlearn outdated information, and integrate new understanding. AI can provide timely updates and resources, but learners must actively engage, evaluate, and adapt their knowledge. Flourishing arises when education emphasizes flexibility, encouraging students to approach change as an opportunity rather than a threat.

Problem-solving becomes the cornerstone of meaningful learning. Students must tackle complex, real-world challenges, combining critical thinking, creativity, and collaboration. AI can simulate scenarios, suggest solutions, and analyze outcomes, but humans must identify problems, make judgments, and apply solutions ethically. Flourishing occurs when students are trained to navigate ambiguity, develop strategies, and take initiative in diverse contexts.

Inquiry-based learning is central to this educational shift. Students are encouraged to ask questions, explore possibilities, and develop independent lines of reasoning. AI can support inquiry by providing data, highlighting patterns, and suggesting connections, but curiosity must be cultivated intrinsically. Flourishing arises when learners are motivated by genuine inquiry, turning questions into pathways for understanding, innovation, and insight.

Interdisciplinary thinking is essential for navigating complexity. Problems in the modern world—climate change, public health, social inequality—require knowledge from multiple domains. AI can assist by integrating information across disciplines, but humans must interpret, synthesize, and apply insights responsibly. Flourishing occurs when students learn to connect ideas across fields, combining diverse perspectives to develop holistic solutions.

Technology reshapes not only how students access information but also how they interact with knowledge. AI-powered tools can analyze student learning patterns, suggest resources, and provide personalized feedback. While this enhances efficiency and engagement, humans must ensure that students retain the ability to think independently, question sources, and synthesize information critically. Flourishing arises when learners balance the convenience of AI with active intellectual engagement, ensuring knowledge is internalized and applied thoughtfully.

Assessment in this new educational model moves beyond rote testing. Traditional exams measure memory more than understanding or creativity. AI can provide real-time analytics on performance, adaptive assessments, and insights into cognitive patterns, but human educators remain essential for evaluating nuance, ethical judgment, and originality. Flourishing occurs when evaluation emphasizes skill development, problem-solving, and innovation rather than simple recall.

Digital literacy is no longer optional—it is central to effective learning. Students must understand how AI works, recognize bias, verify sources, and navigate ethical challenges in digital spaces. Humans must teach learners to critically engage with AI outputs, questioning assumptions, and integrating multiple perspectives. Flourishing arises when students are equipped to use AI intelligently, ethically, and creatively to expand understanding and solve complex problems.

Emotional and social intelligence are also vital. Collaboration, empathy, and communication skills cannot be replaced by AI. While technology can facilitate interactions across distances, humans must learn to navigate interpersonal dynamics, resolve conflict, and understand emotional context. Flourishing occurs when education nurtures both cognitive skills and human relational abilities, producing well-rounded, adaptable learners.

Education beyond memorization requires fostering curiosity as a lifelong habit. When students are encouraged to explore, ask questions, and connect ideas across subjects, learning becomes an engaging, dynamic process. AI can support this by suggesting new areas of inquiry, highlighting patterns, or simulating complex systems, but genuine curiosity must originate within the learner. Flourishing arises when students develop intrinsic motivation to discover, experiment, and expand their understanding continuously.

Creativity and innovation must be nurtured alongside critical thinking. Students need opportunities to experiment, make mistakes, and approach problems in unconventional ways. AI can enhance creative exploration by providing tools, generating simulations, or modeling alternatives, but imaginative leaps are uniquely human. Flourishing occurs when education balances structured guidance with freedom for experimentation, enabling students to think boldly and independently.

Ethical reasoning and civic awareness are essential components of modern learning. Students must understand the social and moral implications of their knowledge and actions, particularly when interacting with AI and other technologies. Teachers can guide discussions on fairness, bias, and responsibility, while AI can provide examples and simulate consequences. Flourishing arises when students combine knowledge, creativity, and ethical reflection to navigate complex societal challenges.

Collaboration and global awareness are increasingly important. Students must develop the skills to work across cultures, disciplines, and perspectives, leveraging diversity to generate innovative solutions. AI can connect learners worldwide and provide insights into different cultural contexts, but humans must practice empathy, negotiation, and communication. Flourishing occurs when students integrate collaborative skills with ethical judgment and creative thinking to thrive in an interconnected world.

Concluding the discussion on education beyond memorization, it is evident that preparing students for the AI era requires a fundamental shift in approach. Memorizing facts is no longer the primary goal; instead, education must emphasize critical thinking, creativity, ethical reasoning, and adaptability. Flourishing arises when students are empowered to navigate abundant information, apply knowledge responsibly, and develop the uniquely human skills that machines cannot replicate.

Teachers and mentors play a pivotal role in this transformation. While AI can personalize learning, provide instant feedback, and suggest pathways, humans are essential for guiding judgment, fostering curiosity, and nurturing moral and social development. Flourishing occurs when education combines technological tools with human guidance, ensuring students grow intellectually, emotionally, and ethically.

Assessment should prioritize understanding and application over rote recall. Projects, collaborative problem-solving, and real-world challenges measure the skills that matter most in an AI-enhanced world. AI can assist in providing insights, analyzing outcomes, and suggesting improvements, but humans must interpret and guide the learning process. Flourishing arises when evaluation reinforces growth, creativity, and ethical thinking rather than simply testing memory.

Ethical and digital literacy must be cultivated alongside cognitive skills. Students need the ability to evaluate sources, detect bias, understand algorithmic influence, and make responsible decisions. AI can support learning, but judgment, reflection, and ethical reasoning remain human responsibilities. Flourishing occurs when learners develop both technological competence and moral awareness, preparing them for a world where AI shapes knowledge, work, and society.

Finally, education should inspire lifelong curiosity and courage. Students must be encouraged to explore boldly, question assumptions, and engage with uncertainty creatively and responsibly. Flourishing arises when learning is viewed as an ongoing journey of discovery, guided by human insight, creativity, and ethical reflection, ensuring that knowledge and wisdom grow hand in hand in the age of AI.

Inspirational Quote:

"Education is no longer about filling minds with facts, but about cultivating wisdom, curiosity, and courage to navigate a world of infinite information." — Anonymous

Chapter . 32

Identity Formation in an Algorithmic Mirror

Young people will define themselves in constant dialogue with machines.

The digital age has created a new environment for identity formation, one in which algorithms play a central role in shaping self-perception. Social media feeds, recommendation systems, and AI-curated content subtly influence how young people see themselves, their interests, and their place in the world. Flourishing arises when adolescents are guided to critically engage with these digital mirrors, developing a stable sense of self that integrates both human and technological influence.

Algorithms often create feedback loops that reinforce existing behaviors and preferences. A young person interested in music, fashion, or gaming may find their feed increasingly focused on that niche, narrowing exposure to new ideas and perspectives. Humans must teach children to recognize these patterns, question the limitations of algorithmic influence, and seek diverse experiences. Flourishing occurs when young people balance algorithmic guidance with exploration and self-directed growth.

Identity is also shaped by comparison. Social media platforms amplify the visibility of curated lives, achievements, and appearances, often creating unrealistic standards. AI may accentuate these effects by predicting content that maximizes engagement, not well-being. Humans must guide adolescents to understand the difference between curated digital personas and authentic life experiences. Flourishing arises when young people cultivate self-awareness, resilience, and critical reflection in the face of algorithmically amplified comparison.

Autonomy in identity formation is crucial. While AI provides suggestions, trends, and social validation, young people must learn to make conscious choices about values, behaviors, and self-expression. This requires reflection, guidance, and support from mentors, educators, and peers. Flourishing occurs when adolescents exercise agency, integrating technology as a tool for self-discovery rather than allowing it to dictate their identity.

Peer influence is magnified in algorithmic environments. AI-driven platforms often highlight content that aligns with popular trends or social consensus, subtly shaping preferences, opinions, and behaviors. Young people may feel pressure to conform to digital norms, risking authenticity in favor of approval. Humans must guide adolescents to evaluate peer influence critically and make choices that reflect personal values and individuality. Flourishing occurs when young people maintain self-direction while engaging meaningfully with social and digital communities.

Self-expression in an AI-mediated world requires careful navigation. Digital tools allow creative output, from music and visual art to writing and coding, but algorithms often prioritize popularity over substance. Young people must learn to balance the desire for recognition with authenticity and integrity. Flourishing arises when adolescents use AI creatively to explore identity, develop skills, and communicate meaningfully rather than simply seeking validation.

The concept of identity itself is evolving. With AI suggesting preferences, curating content, and even generating virtual companions, young people must integrate technological feedback into a coherent sense of self. Reflection, mentorship, and social support are essential to distinguish personal values from algorithmic influence. Flourishing occurs when adolescents critically assess the role of technology in shaping identity, cultivating a sense of self that is informed, empowered, and authentic.

Digital literacy is foundational for healthy identity formation. Young people must understand how algorithms work, recognize bias, and interpret AI-generated insights critically. Education should focus on building skills for discerning information, resisting manipulation, and navigating digital spaces responsibly. Flourishing arises when adolescents combine digital competence with self-awareness and ethical judgment, enabling informed engagement with AI-mediated environments.

Algorithmic feedback loops can subtly influence self-concept over time. A young person's preferences, behaviors, and interactions are continuously analyzed and reinforced by AI, shaping their perception of what is "normal" or desirable. Humans must teach adolescents to question these patterns, reflect on their choices, and retain agency over their evolving identity. Flourishing arises when young people balance algorithmic guidance with independent thought and self-directed growth.

AI-generated comparisons can affect self-esteem and social confidence. Machine-curated content often highlights ideals, trends, and achievements, creating pressure to measure up. Young people must be supported in understanding that digital representations are often curated, incomplete, or exaggerated. Flourishing occurs when adolescents cultivate self-compassion, realistic expectations, and a secure sense of identity that is resilient to external validation.

Exploration remains critical for healthy identity formation. Adolescents should be encouraged to engage in diverse experiences—intellectual, social, and creative—beyond the confines of algorithmic suggestion. AI can provide access to new ideas and perspectives, but humans must scaffold experiences that challenge assumptions and broaden horizons. Flourishing arises when curiosity drives exploration, allowing young people to construct a rich, multifaceted identity.

Social and ethical reflection are intertwined with digital identity. Decisions about posting, sharing, and interacting online shape not only personal reputation but also the social environment of peers. Adolescents must develop an awareness of the consequences of their digital actions, learning to navigate ethical dilemmas thoughtfully. Flourishing occurs when young people combine self-awareness, ethical judgment, and critical thinking to participate responsibly in digital communities.

Identity formation in an algorithmic world also requires cultivating self-authorship. Adolescents must learn to define who they are based on reflection, values, and aspirations, rather than solely on digital validation or algorithmic cues. AI can suggest pathways, highlight trends, and provide feedback, but the ultimate authority over identity must remain human. Flourishing occurs when young people actively shape their sense of self with intention, integrating technology as a supportive tool rather than a determinant.

Digital communities offer both opportunities and challenges for self-expression. Online platforms enable connection, collaboration, and creativity, but they can also expose young people to comparison, judgment, or toxic social dynamics. Humans must help adolescents navigate these spaces thoughtfully, fostering autonomy, resilience, and ethical engagement. Flourishing arises when digital interaction complements real-world identity development, enhancing growth without undermining self-worth.

Reflection and mindfulness are crucial tools for navigating AI-mediated influence. Adolescents must practice observing their thoughts, feelings, and responses to digital stimuli, learning to separate external pressures from internal values. AI can support self-reflection through journaling prompts, behavior tracking, or pattern recognition, but insight and interpretation remain human. Flourishing occurs when reflection cultivates awareness, guiding choices and reinforcing authentic identity.

Encouraging diversity of experience is also critical. Adolescents need exposure to multiple perspectives, cultures, and challenges to develop a resilient and expansive sense of self. AI can facilitate access to global ideas, but humans must guide interpretation and integration. Flourishing arises when young people engage broadly, synthesizing digital and real-world experiences to form identity that is informed, flexible, and enriched.

Concluding the discussion on identity formation in an algorithmic mirror, it is evident that AI profoundly influences how young people perceive themselves, their choices, and their potential. Algorithms can guide, suggest, and reflect behaviors, but they cannot replace critical reflection, ethical reasoning, or authentic self-expression. Flourishing arises when adolescents engage with AI thoughtfully, maintaining autonomy, resilience, and moral awareness while integrating technological feedback into their evolving sense of self.

Guidance from parents, educators, and mentors is essential. Adults provide context, perspective, and support, helping adolescents distinguish between algorithmically amplified signals and authentic personal values. Flourishing occurs when young people receive mentorship that encourages reflection, self-discovery, and responsible decision-making in a digital environment.

Critical thinking and digital literacy are foundational. Adolescents must understand how algorithms work, recognize bias, and interpret digital input thoughtfully. AI can provide insights, track patterns, and suggest pathways, but judgment and discernment remain inherently human. Flourishing arises when young people combine digital competence with self-awareness, ethical reflection, and curiosity.

Resilience is also key. The constant feedback loops of social media, AI-curated content, and virtual interaction can challenge self-esteem and identity stability. Young people must develop the skills to manage pressure, navigate comparison, and maintain confidence in their intrinsic worth. Flourishing occurs when adolescents cultivate inner stability while leveraging AI as a tool for growth and exploration.

Finally, identity formation in an AI-rich world is an opportunity for empowerment. By learning to navigate algorithms thoughtfully, adolescents can integrate curiosity, creativity, courage, and ethical awareness into their personal development. Flourishing arises when technology enhances rather than dictates identity, allowing the next generation to emerge as self-aware, resilient, and responsible contributors to a rapidly evolving world.

Inspirational Quote:

"We do not become who we are by following algorithms, but by reflecting, choosing, and acting with curiosity, courage, and conscience." — Anonymous

Chapter . 33

Digital Companions and Synthetic Relationships

AI friends and mentors raise new emotional and ethical questions.

As AI becomes increasingly sophisticated, children and adults alike are forming bonds with digital companions. From chatbots and virtual pets to AI mentors, these relationships can provide emotional support, guidance, and companionship. While these interactions offer benefits, they also raise questions about attachment, social development, and the nature of human connection. Flourishing arises when humans integrate digital relationships thoughtfully, balancing the advantages of AI companionship with the irreplaceable value of human interaction.

AI companions can enhance learning and personal growth. Virtual mentors can provide personalized feedback, encourage curiosity, and model problem-solving strategies. These interactions can increase confidence, motivation, and skill acquisition. Yet humans must ensure that these digital relationships supplement rather than replace human mentorship, guidance, and emotional support. Flourishing occurs when AI companionship amplifies human development without diminishing authentic interpersonal bonds.

Emotional development is influenced by interactions with AI. Children may form attachments, seek comfort, or share thoughts with AI entities, sometimes in ways similar to human relationships. While AI can simulate empathy and responsiveness, it lacks consciousness, genuine emotional experience, and moral understanding. Humans must help children interpret AI responses critically and maintain a distinction between synthetic and human relationships. Flourishing arises when emotional bonds with AI support learning and growth without compromising human empathy or connection.

Ethical considerations are central. AI companions collect data, influence behavior, and can reinforce patterns based on design choices. Adolescents and adults must understand privacy, consent, and the impact of AI on decision-making and emotional well-being. Flourishing occurs when humans are equipped to navigate these ethical questions, ensuring digital relationships are safe, intentional, and aligned with personal values.

Digital companions can serve as safe spaces for experimentation and learning. Adolescents, in particular, may practice social interactions, conflict resolution, and emotional expression with AI before applying these skills in real-world contexts. AI provides predictable responses, allowing repeated practice and gradual skill-building. Flourishing occurs when these interactions strengthen confidence, emotional intelligence, and ethical reasoning, without replacing genuine human engagement.

AI friendships also pose challenges for authenticity. Since digital companions are programmed to respond in ways that are often agreeable or adaptive, young people may experience a sense of validation without challenge. This can inadvertently reinforce avoidance of difficult conversations or critical reflection. Humans must encourage children and adults to maintain human relationships that involve negotiation, compromise, and authentic feedback. Flourishing arises when AI interactions complement, rather than diminish, the development of real-world relational skills.

Trust is a complex element in synthetic relationships. Humans must differentiate between programmed responses and genuine understanding, recognizing the limitations of AI companionship. Overreliance can lead to distorted expectations of real-world relationships, where nuance, unpredictability, and emotional depth are intrinsic. Flourishing occurs when humans develop discernment, understanding the difference between artificial support and the richness of human connection.

The role of AI in caregiving and mentorship is also expanding. Digital companions can assist in monitoring well-being, encouraging positive habits, and providing guidance. While these tools offer practical benefits, humans must ensure that ethical judgment, empathy, and accountability remain rooted in human oversight. Flourishing arises when AI aids caregiving and mentoring without replacing the human presence that nurtures trust, understanding, and emotional growth.

Digital companionship introduces new dimensions of social learning. AI can simulate diverse personalities, perspectives, and problem-solving approaches, allowing young people to practice empathy and perspective-taking in a controlled environment. These experiences can broaden understanding and enhance adaptability. Flourishing occurs when AI interactions are used to expand relational skills, preparing humans to engage thoughtfully and empathetically with others in the real world.

Boundaries are essential in AI-human relationships. Unlike human friends, digital companions cannot truly consent, experience emotion, or form reciprocal bonds. Adolescents must learn to navigate attachment carefully, understanding the limits of AI companionship and the importance of human connection. Flourishing arises when young people develop a balanced relational ecosystem, where AI complements rather than substitutes authentic interaction.

Ethical reflection extends to the design and use of AI companions. Users should consider whose values are embedded in these systems, how data is collected, and the potential influence on thoughts, behaviors, and identity. Adults can model critical inquiry, helping young people question assumptions and engage with technology responsibly. Flourishing occurs when users approach digital companions with both curiosity and ethical awareness, fostering conscious and deliberate engagement.

AI companions can also support creativity and self-expression. Digital entities can inspire storytelling, artistic creation, or exploratory dialogue, providing safe spaces to experiment with ideas. Humans must ensure that these tools encourage originality without constraining thought to algorithmically suggested paths. Flourishing arises when creativity is nurtured through AI, yet remains firmly grounded in human imagination and intention.

Digital companions also raise questions about attachment and dependency. While AI can provide comfort, guidance, and a sense of presence, overreliance may hinder the development of resilience, patience, and human empathy. Humans must learn to set boundaries for themselves and for children interacting with AI, ensuring that emotional needs are met primarily through human connection. Flourishing arises when AI supports emotional growth without replacing the richness and unpredictability of real-world relationships.

Privacy and security are critical considerations in synthetic relationships. Digital companions collect and process personal information, which can influence behavior, decision-making, and identity. Adolescents and adults must develop awareness of data privacy, ethical design, and the implications of sharing personal experiences with AI. Flourishing occurs when humans engage with digital companions thoughtfully, protecting autonomy and exercising informed judgment about the digital traces they leave behind.

Mentorship through AI has potential, but human guidance remains irreplaceable. AI can suggest strategies, model solutions, and offer encouragement, yet it lacks moral judgment, intuition, and emotional nuance. Humans must interpret and contextualize AI feedback, ensuring that learning and development remain grounded in ethical and experiential wisdom. Flourishing arises when AI enhances mentorship without replacing the human presence that provides accountability, empathy, and inspiration.

Synthetic relationships also invite reflection on authenticity. Engaging with AI encourages users to question what constitutes a "real" relationship, a "meaningful" interaction, and a "trusted" companion. Humans must cultivate awareness that while AI can simulate connection, genuine emotional reciprocity exists only among conscious beings. Flourishing occurs when individuals integrate AI into their social and emotional lives consciously, valuing authentic human relationships while leveraging digital companions as tools for growth.

Concluding the discussion on digital companions and synthetic relationships, it is clear that AI can both enrich and complicate human emotional life. These interactions offer opportunities for learning, creativity, and support, yet they require careful guidance, ethical reflection, and awareness of their limitations. Flourishing arises when humans use AI companionship intentionally, complementing human relationships rather than replacing them.

Adults play a pivotal role in shaping the experience of synthetic relationships. Parents, educators, and mentors provide context, boundaries, and perspective, helping young people interpret AI interactions critically. They can guide adolescents to understand the difference between programmed responsiveness and authentic emotion. Flourishing occurs when guidance ensures that AI companionship supports growth, empathy, and self-awareness without undermining human connection.

Ethical literacy is essential. Users must consider the design, data collection, and behavioral influence of AI companions. Critical reflection and informed decision-making protect autonomy, privacy, and emotional well-being. Flourishing arises when humans engage thoughtfully with AI, integrating ethical awareness into their interactions and ensuring that relationships—digital or human—are intentional and enriching.

Creativity and exploration benefit from synthetic companionship. AI can stimulate imagination, provide feedback, and simulate diverse scenarios, fostering skill-building and confidence. Yet, originality remains a human trait. Flourishing occurs when AI serves as a tool to amplify creativity and learning while humans retain ownership of ideas, expression, and ethical judgment.

Finally, the central lesson is balance. AI companionship should enhance development without replacing the unpredictability, empathy, and moral complexity inherent in human relationships. Flourishing arises when humans cultivate curiosity, emotional intelligence, ethical reasoning, and reflection, integrating AI thoughtfully into lives that remain anchored in authentic human connection.

Inspirational Quote:

"*AI companions can guide and inspire, but the heart of human connection—empathy, trust, and love—remains irreplaceable.*" — Anonymous

Chapter . 34

Teaching Wisdom, Not Just Knowledge

The future demands character development as much as technical skill.

Education in the age of AI must go beyond transmitting knowledge to cultivating wisdom—the ability to apply knowledge ethically, reflectively, and insightfully. While machines can process facts, analyze patterns, and provide answers, they cannot teach discernment, judgment, or moral reasoning. Flourishing arises when humans prioritize character, ethical reasoning, and reflective thinking alongside technical skill, preparing students to navigate a complex, AI-augmented world.

Knowledge without judgment is insufficient. Students may acquire vast amounts of information, but without the ability to evaluate its relevance, accuracy, or ethical implications, learning is incomplete. AI can provide instant facts and simulations, but humans must teach students how to discern, question, and apply knowledge thoughtfully. Flourishing occurs when education integrates knowledge with reflection, critical thinking, and ethical consideration.

Empathy and social understanding are central to wisdom. The ability to navigate relationships, consider multiple perspectives, and act compassionately is uniquely human. AI can model social dynamics, simulate outcomes, or highlight patterns, but emotional intelligence and ethical insight remain human capacities. Flourishing arises when education cultivates empathy, perspective-taking, and moral awareness, ensuring students can act wisely in social and professional contexts.

Decision-making skills must also be nurtured. Students should learn to anticipate consequences, weigh competing priorities, and make choices aligned with values. AI can assist by modeling scenarios and predicting outcomes, but final judgment rests with humans. Flourishing occurs when learners develop the confidence and competence to make informed, ethical, and thoughtful decisions in a world of rapid technological change.

Teaching wisdom also involves encouraging long-term thinking. In an age of instant gratification and algorithmically optimized choices, students must learn to consider consequences beyond immediate outcomes. AI can model projections, simulate risks, and visualize scenarios, but humans must evaluate meaning, values, and ethical impact over time. Flourishing arises when learners integrate foresight, responsibility, and ethical reflection into their decision-making processes.

Ethical dilemmas should be central to learning. Students can explore scenarios where choices affect others, challenge fairness, or raise moral questions. AI can provide simulated experiences, highlight outcomes, or present alternative perspectives, but humans guide the interpretation and moral reasoning. Flourishing occurs when students develop a framework for evaluating decisions ethically, preparing them to act responsibly in complex social, technological, and environmental contexts.

Creativity and innovation are inseparable from wisdom. Knowledge without imagination is limited, but imagination without judgment can be reckless. AI can assist by generating ideas, modeling patterns, or suggesting alternatives, but the discernment to choose, refine, and apply ideas wisely remains human. Flourishing arises when learners are trained to balance originality with prudence, exploring boldly while remaining grounded in ethics and reflection.

Communication and collaboration are also essential components of wisdom. Students must learn to listen, negotiate, and express themselves with clarity and empathy. AI can facilitate connection, provide translation, and analyze discourse patterns, but human interaction is necessary for understanding nuance, emotion, and intent. Flourishing occurs when learners cultivate relational skills, integrating knowledge, reflection, and empathy into meaningful collaboration.

Wisdom education also requires integrating self-awareness. Students must understand their strengths, weaknesses, biases, and motivations, using this understanding to guide choices and behavior. AI can assist by providing feedback, tracking patterns, and suggesting areas for improvement, but the capacity for introspection, ethical judgment, and self-regulation is inherently human. Flourishing arises when learners cultivate deep self-knowledge, enabling them to navigate both personal and societal challenges with clarity and integrity.

Mentorship and modeling play critical roles in teaching wisdom. Students benefit from observing adults who demonstrate ethical decision-making, empathy, and reflective practice. AI can simulate scenarios, but cannot embody moral authority or lived experience. Flourishing occurs when human role models, alongside AI tools, help learners integrate knowledge with virtue, judgment, and discernment, forming the foundation for wise action.

Interdisciplinary understanding enhances wisdom. Ethical, emotional, and practical considerations often span multiple domains—science, technology, humanities, and social systems. AI can provide cross-disciplinary analysis, identify patterns, and suggest connections, but humans must synthesize insights to make ethically informed, practical decisions. Flourishing arises when learners integrate diverse perspectives into coherent understanding, applying wisdom to complex, interconnected challenges.

The cultivation of moral imagination is another key component. Students should envision possible futures, empathize with stakeholders, and anticipate consequences, fostering foresight and responsibility. AI can simulate scenarios and visualize outcomes, yet imagination grounded in ethical reasoning remains human. Flourishing occurs when learners develop the ability to envision the impact of their actions and act with consideration for others, society, and the environment.

Teaching wisdom requires emphasizing purpose and meaning alongside skill acquisition. Knowledge and technical competence alone are insufficient if they are not guided by values and a sense of responsibility. AI can support exploration of career paths, societal challenges, and personal interests, but it cannot provide the moral framework or life perspective necessary to act wisely. Flourishing arises when education cultivates both competence and conscience, guiding learners to align action with purpose and ethical principles.

Students must also learn to navigate uncertainty and complexity. AI can generate predictions and solutions, yet the world is inherently unpredictable and morally ambiguous. Humans must teach learners to weigh incomplete information, adapt to changing circumstances, and make choices aligned with ethical reflection. Flourishing occurs when students integrate judgment, creativity, and resilience, using AI as a tool rather than a crutch in navigating complexity.

Collaboration is central to the cultivation of wisdom. Ethical and practical challenges often require collective insight, negotiation, and compromise. AI can facilitate communication, provide data-driven suggestions, and highlight potential solutions, but human judgment, empathy, and shared responsibility are essential for meaningful outcomes. Flourishing arises when learners are equipped to engage thoughtfully and responsibly with others, using AI to augment collaboration rather than replace human interaction.

Reflective practice must be habitual. Students should regularly examine the consequences of their choices, the alignment of their actions with values, and their impact on others. AI can provide insights, track trends, and model scenarios, but interpretation and moral deliberation remain human responsibilities. Flourishing occurs when learners develop the ability to pause, evaluate, and adjust their behavior in service of wisdom, integrating insight, experience, and ethical reflection.

Concluding the discussion on teaching wisdom, it is clear that education must focus on character, judgment, and ethical reflection as much as knowledge and skill. While AI can provide information, guidance, and simulation, the human capacities for discernment, empathy, and moral reasoning remain irreplaceable. Flourishing arises when learners are prepared to apply knowledge thoughtfully, act ethically, and navigate complexity with both insight and integrity.

Mentorship and human modeling are central to this endeavor. Teachers, parents, and mentors provide examples of wise action, demonstrating how knowledge, ethical reflection, and courage combine in real-life decision-making. AI can support this process through data, scenario modeling, and feedback, but the moral and relational guidance of humans is indispensable. Flourishing occurs when learners internalize these lessons, integrating knowledge with character development and ethical judgment.

Ethical literacy and critical thinking are inseparable from wisdom. Students must evaluate information, reflect on implications, and consider broader societal consequences. AI can augment understanding by offering analysis and predictions, but it cannot instill conscience, perspective, or judgment. Flourishing arises when learners combine analytical insight with moral reasoning, applying wisdom in both personal and societal contexts.

Creativity, curiosity, and reflection complete the foundation for wisdom. Learners must engage actively with ideas, explore new possibilities, and evaluate choices critically. AI can inspire, simulate, and provide tools for exploration, but imagination, ethical reflection, and thoughtful decision-making remain human domains. Flourishing occurs when students integrate these capabilities, applying knowledge with insight, compassion, and ethical discernment.

Finally, the ultimate goal of teaching wisdom is human flourishing itself. By cultivating judgment, empathy, courage, and reflection alongside knowledge and skill, education prepares learners not only to navigate a technologically advanced world but also to enhance it ethically and responsibly. Flourishing arises when humans are empowered to act wisely, creatively, and ethically, ensuring that AI serves as a tool for the betterment of humanity rather than a substitute for human judgment.

Author Reflection:

Writing this chapter emphasized that wisdom is the bridge between knowledge and meaningful action. In an age where AI provides instant answers, the human capacity for reflection, ethical judgment, and courageous action becomes more critical than ever. True flourishing occurs when learners integrate knowledge, character, and discernment, using AI as a tool to enhance—not replace—human wisdom.

Inspirational Quote:

"*Knowledge equips the mind, but wisdom guides the heart and shapes the world.*" — Anonymous

Part VIII — Choosing a Flourishing Future

Chapter . 35

Designing AI for Human Good

Technology reflects the values of those who build it.

AI is not neutral; it reflects the intentions, values, and assumptions of its creators. Every algorithm, dataset, and design decision embodies choices about what is prioritized, what is measured, and whose interests are served. Flourishing arises when humans intentionally design AI to support human well-being, equity, and ethical outcomes rather than unconsciously reproducing bias or harm.

Human-centered design is central to creating beneficial AI. Designers must prioritize human needs, ethical standards, and societal impact at every stage. This involves considering accessibility, fairness, transparency, and long-term consequences. Flourishing occurs when AI development is guided by empathy and foresight, ensuring that technology amplifies human potential rather than exacerbating inequality or injustice.

Ethical frameworks provide critical guidance in AI design. Principles such as fairness, accountability, transparency, and privacy help shape systems that respect human rights and dignity. AI should be evaluated not only for performance but also for its impact on individuals, communities, and societies. Flourishing arises when developers integrate ethical reflection into every phase of AI creation, aligning innovation with the collective good.

Stakeholder engagement enhances the design of human-centered AI. Involving diverse perspectives—from end users to affected communities—ensures that AI systems address real needs and avoid unintentional harm. AI that reflects only the worldview of its developers risks reinforcing inequities and blind spots. Flourishing occurs when technology is co-created through inclusive, participatory processes that honor the voices of all stakeholders.

Designing AI for human good requires anticipating unintended consequences. Even well-intentioned systems can produce outcomes that harm individuals or communities if social, cultural, or ethical dimensions are overlooked. Developers must engage in scenario planning, ethical impact assessments, and interdisciplinary collaboration to foresee risks. Flourishing occurs when AI is designed proactively, with attention to potential pitfalls and mechanisms for correction.

Transparency is crucial in fostering trust between humans and AI. Users must understand how AI systems make decisions, what data they rely on, and where limitations exist. Black-box models may perform effectively, but they risk alienating users and undermining accountability. Flourishing arises when AI systems are explainable and interpretable, enabling humans to make informed choices and hold technology accountable.

Bias mitigation is a key responsibility of AI designers. Historical inequities and structural discrimination can be encoded into algorithms if unchecked. Careful curation of training data, fairness auditing, and ongoing monitoring are necessary to prevent AI from reinforcing societal inequities. Flourishing occurs when designers intentionally detect, address, and prevent bias, ensuring that AI serves as a tool for inclusion and equity.

Collaboration between humans and AI developers strengthens outcomes. Ethical, technical, and social perspectives must intersect in the design process. Engineers, ethicists, sociologists, and community representatives contribute to creating AI that aligns with human values. Flourishing arises when diverse expertise informs design, producing technology that benefits society while respecting individual rights and dignity.

Designing AI for human good also requires prioritizing inclusivity. Systems must serve people of all backgrounds, abilities, and socioeconomic circumstances. Neglecting marginalized or underrepresented groups risks deepening inequality and perpetuating exclusion. Flourishing arises when AI design consciously incorporates diverse perspectives, ensuring equitable access and benefit for all members of society.

Ethical decision-making in AI involves trade-offs. Designers often navigate tensions between efficiency, privacy, fairness, and usability. No system is perfect, but explicit reflection on ethical priorities and transparent communication about choices mitigates harm. Flourishing occurs when humans consciously weigh trade-offs, prioritizing outcomes that enhance well-being and uphold societal values.

Education and training are integral to responsible AI development. Engineers, designers, and decision-makers must be equipped with ethical reasoning, social awareness, and technical expertise. Understanding the societal consequences of AI empowers creators to design systems that reflect human-centered principles. Flourishing arises when education fosters the skills, judgment, and awareness necessary to build technology aligned with the common good.

Community feedback loops strengthen AI systems over time. Users provide insights, identify issues, and suggest improvements that designers may not anticipate. Iterative design processes, informed by lived experience, make AI more responsive, safe, and beneficial. Flourishing occurs when AI evolves in dialogue with the people it serves, combining technological innovation with social insight.

Human values must be at the core of AI development. Designers must articulate the principles they wish to uphold—fairness, equity, empathy, environmental stewardship, and social benefit—and ensure that these values guide every stage of the process. AI should not simply optimize for efficiency or profit; it must reflect ethical priorities that support human flourishing. Flourishing occurs when technology serves as a mirror of humanity's highest aspirations rather than a reflection of short-term gain.

Interdisciplinary collaboration is essential to embed human-centered values into AI. Engineers alone cannot anticipate societal consequences; ethicists, social scientists, educators, and community stakeholders must contribute to design, testing, and deployment. This collective approach ensures that AI systems respect diverse needs, cultural contexts, and human dignity. Flourishing arises when AI is developed through inclusive processes that integrate multiple perspectives into the architecture of innovation.

Transparency in design fosters trust and accountability. Users must understand how AI reaches decisions, which data informs predictions, and where errors may arise. Transparency empowers people to engage critically with AI, challenge potential harms, and make informed choices. Flourishing occurs when AI is intelligible, explainable, and accessible, allowing humans to navigate technological systems with awareness and control.

Monitoring and adaptation are continuous responsibilities. AI systems are not static; they evolve with data, user interaction, and environmental changes. Designers must implement mechanisms for oversight, feedback, and iterative improvement. Flourishing arises when AI is guided by ongoing human supervision, ensuring that ethical principles remain central as technology adapts and grows.

Human values must be at the core of AI development. Designers must articulate the principles they wish to uphold—fairness, equity, empathy, environmental stewardship, and social benefit—and ensure that these values guide every stage of the process. AI should not simply optimize for efficiency or profit; it must reflect ethical priorities that support human flourishing. Flourishing occurs when technology serves as a mirror of humanity's highest aspirations rather than a reflection of short-term gain.

Interdisciplinary collaboration is essential to embed human-centered values into AI. Engineers alone cannot anticipate societal consequences; ethicists, social scientists, educators, and community stakeholders must contribute to design, testing, and deployment. This collective approach ensures that AI systems respect diverse needs, cultural contexts, and human dignity. Flourishing arises when AI is developed through inclusive processes that integrate multiple perspectives into the architecture of innovation.

Transparency in design fosters trust and accountability. Users must understand how AI reaches decisions, which data informs predictions, and where errors may arise. Transparency empowers people to engage critically with AI, challenge potential harms, and make informed choices. Flourishing occurs when AI is intelligible, explainable, and accessible, allowing humans to navigate technological systems with awareness and control.

Monitoring and adaptation are continuous responsibilities. AI systems are not static; they evolve with data, user interaction, and environmental changes. Designers must implement mechanisms for oversight, feedback, and iterative improvement. Flourishing arises when AI is guided by ongoing human supervision, ensuring that ethical principles remain central as technology adapts and grows.

Finally, designing AI for human good requires humility. Developers must acknowledge uncertainty, limitations, and the potential for unintended consequences. Rather than assuming perfection, humans must approach AI with caution, reflection, and a commitment to continuous learning. Flourishing occurs when humility informs innovation, guiding designers to prioritize ethical responsibility, societal benefit, and the enduring well-being of humanity.

Concluding the discussion on designing AI for human good, it is clear that technology is a reflection of human values and intentions. Ethical design requires foresight, inclusivity, transparency, and ongoing accountability. Flourishing arises when AI is created deliberately to enhance human well-being, amplify human potential, and mitigate harm, rather than reproducing inequities or prioritizing efficiency over ethics.

Human oversight and reflection are essential throughout the lifecycle of AI. Designers, policymakers, and users must remain vigilant, ensuring that AI aligns with societal values and human needs. Iterative evaluation, community engagement, and ethical review help identify blind spots and correct unintended consequences. Flourishing occurs when technology evolves under thoughtful human stewardship, creating systems that serve people responsibly and equitably.

Education and interdisciplinary collaboration strengthen ethical AI development. Engineers, ethicists, social scientists, and affected communities contribute diverse perspectives, ensuring that AI is grounded in human understanding and moral reflection. Flourishing arises when technological innovation is inseparable from ethical awareness, cultural insight, and social responsibility.

Inclusivity and fairness are non-negotiable principles. AI must serve all individuals, regardless of background, socioeconomic status, or ability, while mitigating bias and promoting equity. Humans must design systems that amplify opportunity rather than perpetuate disadvantage. Flourishing occurs when AI supports justice, equality, and empowerment for all members of society.

Finally, humility and responsibility guide sustainable AI development. Designers must recognize limitations, anticipate consequences, and accept accountability for the impact of their creations. Flourishing arises when humans approach AI with reflection, courage, and ethical integrity, ensuring that technology contributes positively to the collective human journey.

Chapter . 36

Governance, Responsibility, and Global Cooperation

Unchecked AI power requires coordinated human oversight.

AI is evolving at a pace that outstrips most regulatory and governance frameworks. Without deliberate oversight, powerful AI systems can exacerbate inequality, amplify bias, and destabilize social, economic, and political systems. Flourishing arises when humans implement governance structures that ensure AI is developed and deployed responsibly, ethically, and in alignment with global well-being.

National and international regulation is essential to manage AI's impact. Governments, international organizations, and civil society must collaborate to establish standards, norms, and legal frameworks that protect human rights, privacy, and fairness. Flourishing occurs when oversight mechanisms are proactive, transparent, and enforceable, ensuring that AI serves humanity rather than uncontrolled interests.

Corporate responsibility is equally important. Companies creating AI must be accountable for the social, ethical, and environmental consequences of their systems. Ethical audits, impact assessments, and reporting mechanisms help ensure that business goals do not override societal welfare. Flourishing arises when corporate actors embrace responsibility as integral to innovation, prioritizing human and societal well-being alongside profitability.

Global cooperation addresses the borderless nature of AI. Technology transcends national boundaries, and challenges such as autonomous weapons, global surveillance, and economic disruption require coordinated action. Collaboration between nations, research institutions, and NGOs is critical for preventing misuse and ensuring equitable access to benefits. Flourishing occurs when humanity works collectively to guide AI toward common goals and shared prosperity.

Ethical responsibility in AI governance begins with clear accountability. Developers, organizations, and governments must be answerable for decisions and outcomes related to AI deployment. Without clarity about who is responsible, harm can proliferate unchecked, eroding trust and social cohesion. Flourishing occurs when systems of accountability are robust, transparent, and enforceable, ensuring that ethical responsibility is not optional but embedded.

Risk assessment and mitigation are essential components of responsible AI governance. Potential harms—ranging from bias in decision-making to economic displacement—must be identified proactively, with strategies in place to prevent or minimize negative outcomes. Flourishing arises when stakeholders anticipate challenges, incorporate safeguards, and continually monitor AI's social and ethical impact, rather than reacting only after harm occurs.

International standards and cooperation foster alignment across borders. AI technologies developed in one country can affect individuals and societies globally, making unilateral regulation insufficient. Collaborative frameworks, treaties, and shared principles can help harmonize expectations, prevent misuse, and promote equitable distribution of AI benefits. Flourishing occurs when global collaboration ensures that AI serves humanity collectively rather than privileging a few nations or corporations.

Transparency and explainability are vital for public trust and effective governance. Citizens, regulators, and organizations must understand how AI systems make decisions and the rationale behind outcomes. Black-box systems erode confidence and can conceal bias or abuse. Flourishing arises when AI is intelligible and interpretable, allowing humans to hold developers accountable and make informed decisions about deployment and oversight.

Building resilience into AI governance is essential. Systems must be prepared to respond to unforeseen consequences, technological failures, or malicious use. Contingency planning, stress testing, and ethical scenario modeling strengthen the capacity to mitigate risks before they escalate. Flourishing arises when humans anticipate uncertainty, integrating foresight and precaution into regulatory frameworks to safeguard society and individuals.

Public-private partnerships are key to effective oversight. Collaboration between governments, corporations, academic institutions, and civil society ensures that multiple perspectives inform policy, innovation, and ethical considerations. Such partnerships promote transparency, accountability, and equitable distribution of benefits. Flourishing occurs when diverse actors work together, combining expertise, resources, and ethical commitment to guide AI development responsibly.

Cultural and contextual sensitivity is critical in governance. AI impacts communities differently depending on historical, social, and economic circumstances. Policies that succeed in one context may fail or cause harm in another if local realities are ignored. Flourishing arises when global cooperation is coupled with contextual understanding, ensuring that AI benefits are inclusive and culturally appropriate.

Ethical education complements governance frameworks. Citizens, developers, and policymakers must understand AI's social, economic, and moral implications to engage responsibly. Training in ethics, critical thinking, and digital literacy equips individuals to participate meaningfully in oversight and decision-making. Flourishing occurs when informed, ethical citizens shape AI governance, reinforcing accountability and collective responsibility.

Governance frameworks must also address the economic implications of AI. Automation, labor displacement, and wealth concentration pose risks that require thoughtful policy interventions. Regulations around taxation, labor protections, and equitable access to AI-driven benefits help ensure societal stability. Flourishing occurs when governance balances innovation with social responsibility, ensuring that economic progress does not leave marginalized populations behind.

Ethical oversight should be proactive rather than reactive. Anticipating potential harms—such as surveillance misuse, algorithmic discrimination, or autonomous weaponry—requires foresight, interdisciplinary expertise, and moral reflection. AI governance must not wait for crises to act; preventative measures are far more effective. Flourishing arises when humans embed ethical vigilance into AI policy and deployment from the outset, reducing harm before it occurs.

Global cooperation also enhances innovation. Shared standards, knowledge exchange, and collaborative research can accelerate beneficial AI applications while minimizing risks. Nations that cooperate can ensure that AI serves humanity collectively rather than fueling competition that prioritizes speed over safety or equity. Flourishing occurs when AI development is guided by collaboration, mutual accountability, and shared humanistic goals.

Community engagement strengthens legitimacy and trust. Citizens should participate in discussions, consultations, and decision-making processes around AI deployment. Transparent dialogue fosters public understanding and ensures that policies reflect societal priorities. Flourishing arises when governance structures empower individuals to contribute to the ethical and responsible shaping of AI's role in society.

Concluding the discussion on governance, responsibility, and global cooperation, it is clear that AI's potential can only be realized ethically when humans exercise oversight and collaboration. Unchecked technological power risks exacerbating inequality, bias, and instability, while deliberate, inclusive governance ensures that AI aligns with human values and societal well-being. Flourishing arises when regulation, ethical reflection, and global cooperation work together to guide AI toward positive impact.

Accountability is central to this process. Developers, corporations, governments, and international bodies must take responsibility for AI systems' design, deployment, and outcomes. Clear lines of authority and transparent reporting reinforce ethical use and public trust. Flourishing occurs when humans accept and act upon their responsibility, recognizing that ethical stewardship is integral to technological progress.

Education and public engagement strengthen governance. Informed citizens, empowered to participate in decision-making, contribute to oversight, advocacy, and ethical evaluation of AI systems. Ethical literacy, digital fluency, and critical thinking equip society to navigate complex technological challenges. Flourishing arises when humans collectively shape AI governance, ensuring that policies reflect human needs, rights, and values.

Global coordination is indispensable. The transnational nature of AI requires countries and organizations to align on standards, safeguards, and equitable access to benefits. Cooperation prevents technological arms races, reduces misuse, and ensures shared prosperity. Flourishing occurs when nations and communities work together, embedding collaboration, fairness, and ethical principles into the fabric of AI governance.

Finally, governance is an ongoing responsibility, not a one-time achievement. AI evolves rapidly, and oversight structures must adapt, incorporating feedback, monitoring outcomes, and revising policies in response to new challenges. Flourishing arises when humans maintain vigilance, humility, and ethical commitment, ensuring that AI development enhances society while upholding human dignity and global well-being.

Author Reflection:

Writing this chapter reinforced that AI cannot govern itself; human responsibility is essential. Effective governance requires accountability, education, global cooperation, and ethical foresight. True flourishing arises when humans act collectively to ensure that AI amplifies human potential, fairness, and shared prosperity rather than creating unregulated risk or inequity.

Inspirational Quote:

"*Technology without governance is power without responsibility; only human oversight ensures that AI serves humanity, not the other way around.*" — Anonymous

Chapter . 37

Slowing Down in a World That Accelerates

Human well-being depends on resisting constant optimization.

Modern life, increasingly mediated by AI and automation, moves at a pace that can overwhelm human capacity. Tasks are optimized, decisions accelerated, and information streams continuous, leaving little space for reflection, rest, or deep engagement. Flourishing arises when humans intentionally slow down, creating space to think, feel, and connect amidst the constant acceleration of technology.

Slowing down begins with conscious awareness. Humans must recognize the pressures and impulses created by rapid technological environments and the ways AI encourages instant decision-making and perpetual engagement. Awareness allows individuals to choose when to pause, reflect, and engage thoughtfully rather than reacting compulsively. Flourishing occurs when mindfulness and intentionality guide interaction with both technology and daily life.

The quality of attention is crucial. In a world of notifications, alerts, and algorithmically curated content, deep focus is increasingly rare. Humans must cultivate practices that preserve sustained attention and reduce distraction, such as deliberate time blocks, mindfulness exercises, and periods of disconnection.

Flourishing arises when attention is consciously managed, allowing humans to engage fully with meaningful work, relationships, and self-reflection.

Emotional regulation is another benefit of slowing down. Rapid technological feedback loops can intensify stress, anxiety, and emotional reactivity. Humans need strategies to pause, breathe, and process experiences before responding. Flourishing occurs when slowing down enables emotional clarity, resilience, and thoughtful interaction, fostering both individual well-being and harmonious relationships.

Slowing down is also essential for ethical reflection. In a world dominated by AI-driven recommendations and rapid optimization, decisions can become automated or reactive, bypassing moral consideration. Humans must create pauses to assess consequences, weigh values, and consider broader impact before acting. Flourishing arises when reflection replaces reflexive behavior, ensuring that actions are guided by conscience rather than convenience.

Connection with others deepens when time is unhurried. Relationships require listening, empathy, and presence—qualities that are compromised when life is rushed or mediated by AI-driven efficiency. Slowing down allows space for genuine dialogue, shared experiences, and meaningful engagement. Flourishing occurs when humans prioritize connection, nurturing social bonds that provide support, understanding, and joy.

Integration of mind and body is supported by slowing down. Physical and mental health suffer in environments that encourage constant acceleration, multitasking, and digital immersion. Practices such as mindful movement, meditation, and intentional breaks help restore balance and energy. Flourishing arises when humans honor their physiological and psychological needs, cultivating resilience and vitality alongside productivity.

Slowing down also enhances learning and comprehension. Information is abundant, but deep understanding requires reflection, questioning, and synthesis. AI can provide data and insights, but humans must engage thoughtfully to integrate knowledge meaningfully. Flourishing occurs when slowing down enables comprehension, wisdom, and skill development rather than superficial absorption of facts.

Intentional slowing down cultivates resilience in the face of change. Rapid technological shifts, social transformations, and AI-driven disruptions can create stress, anxiety, and uncertainty. By creating space to pause, humans can process challenges, evaluate options, and respond with clarity rather than panic. Flourishing arises when deliberate reflection allows adaptability and calmness, turning pressure into opportunity for growth.

Creative thinking benefits from unhurried time. Many insights, breakthroughs, and innovations emerge not from constant action but from reflection, incubation, and free exploration. AI can assist by generating ideas and suggestions, but the spark of human originality requires mental space. Flourishing occurs when slowing down provides room for imagination, experimentation, and the emergence of unique solutions.

Slowing down also fosters ethical leadership. Leaders navigating AI-enhanced environments must resist impulsive decisions and consider long-term societal, organizational, and environmental consequences. Reflection, consultation, and deliberation are essential to guide teams, communities, and societies responsibly. Flourishing arises when slowing down enables leaders to act with wisdom, integrity, and foresight.

Attention to personal values and purpose is strengthened by a slower pace. In accelerated systems, people often chase efficiency or external metrics at the expense of meaning. Pausing allows humans to align actions with priorities, evaluate life goals, and maintain moral and emotional integrity. Flourishing occurs when humans consciously integrate purpose into daily decisions, ensuring life is guided by values rather than constant external pressures.

Slowing down encourages mindful interaction with AI. As machines provide instant feedback, recommendations, and optimization, humans can feel compelled to act quickly, often sacrificing reflection. By intentionally pausing, users can critically evaluate AI outputs, question assumptions, and make more thoughtful decisions. Flourishing arises when humans maintain agency and discernment, using AI as a tool rather than succumbing to algorithmic speed.

Digital boundaries are essential in slowing down. Notifications, alerts, and endless streams of information can fragment attention and increase stress. Establishing limits on screen time, AI interactions, and constant connectivity allows humans to reclaim focus and mental clarity. Flourishing occurs when boundaries preserve cognitive and emotional health, enabling deep engagement with both technology and life.

Slowing down also fosters learning from mistakes. In high-speed environments, errors can go unnoticed, unexamined, or repeated. Pausing provides space to analyze outcomes, reflect on decisions, and adjust strategies. Flourishing arises when humans integrate reflection and learning into their responses, cultivating wisdom and adaptability even amid rapid technological change.

Time for reflection strengthens social cohesion and community engagement. Relationships, collaborative problem-solving, and civic participation all benefit from unhurried attention. AI can facilitate communication, but meaningful human connection requires presence, patience, and listening. Flourishing occurs when humans slow down to nurture relationships and build stronger, more resilient communities.

Concluding the discussion on slowing down in a world of acceleration, it is clear that human flourishing requires intentional pauses amidst constant technological speed. AI, automation, and digital connectivity offer efficiency and capability, yet they also pressure humans toward constant action and rapid response. Flourishing arises when humans reclaim time for reflection, creativity, ethical consideration, and connection, balancing technological power with human insight.

Mindfulness and intentionality are essential practices. Slowing down allows humans to observe, reflect, and choose deliberately rather than reacting to external stimuli or algorithmic prompts. Pausing fosters clarity, emotional resilience, and informed decision-making. Flourishing occurs when humans cultivate self-awareness, preserving autonomy and conscious engagement in the face of accelerating systems.

Slowing down nurtures creativity, wisdom, and ethical reflection. Unhurried time allows ideas to incubate, perspectives to broaden, and moral judgments to mature. AI can augment human thought, but insight and ethical action emerge only when humans allow space to think, reflect, and imagine. Flourishing arises when time is protected as a resource for deliberate, thoughtful, and principled action.

Relationships and community benefit from slower rhythms. Connection, empathy, and collaboration require presence and attention, which are diminished in a world of perpetual acceleration. By slowing down, humans can invest fully in meaningful interaction, strengthening social bonds and collective resilience. Flourishing occurs when humans prioritize relationships and community alongside technological productivity.

Finally, slowing down restores alignment with purpose, values, and inner life. Human flourishing is realized when individuals balance action with reflection, efficiency with meaning, and technology with consciousness. Pausing is not a retreat but a strategy for living fully, ethically, and creatively in a rapidly changing world. Flourishing arises when humans harness the power of slowing down to integrate mind, heart, and purpose in an AI-driven society.

Author Reflection:

Writing this chapter emphasized that speed and optimization, while valuable, are not substitutes for reflection, ethics, or presence. Human flourishing depends on creating space for thought, creativity, and connection in a world that constantly accelerates. Slowing down is not a luxury—it is a necessity for living intentionally and ethically amidst AI's rapid expansion.

Inspirational Quote:

"*In the rush of progress, the pause is where wisdom grows, creativity blooms, and humanity thrives.*" — Anonymous

Chapter . 38

Reclaiming Community and Shared Meaning

Flourishing is collective, not individual.

Human flourishing depends on the strength of communities and the shared meanings that bind them. In an AI-driven world, individual achievement is often celebrated over collective well-being, and efficiency can overshadow human connection. Flourishing arises when people prioritize relationships, mutual support, and collective purpose alongside personal growth.

Community provides identity, belonging, and resilience. Humans are social beings, and meaningful connections are essential for emotional, cognitive, and moral development. AI can facilitate interaction, bridge distances, and provide resources, but it cannot replace the richness, empathy, and accountability inherent in human networks. Flourishing occurs when communities cultivate trust, reciprocity, and shared values, integrating technology as a tool rather than a replacement for human connection.

Shared meaning anchors societies. Rituals, stories, and cultural practices convey values and collective identity, guiding ethical behavior and social cohesion. AI can support the preservation and dissemination of cultural knowledge, but it cannot generate authentic meaning. Flourishing arises when humans actively participate in creating, maintaining, and evolving shared narratives that foster belonging and purpose.

Collective problem-solving is enhanced through intentional community engagement. Many challenges amplified by AI—such as economic inequality, ethical dilemmas, and environmental crises—require collaborative solutions. Humans must work together to deliberate, innovate, and implement strategies that balance efficiency with justice. Flourishing occurs when communities harness collective intelligence, empathy, and ethical reasoning to address shared challenges.

Reclaiming community requires intentional design of social spaces, both physical and digital. Neighborhoods, workplaces, schools, and online platforms can either foster connection or isolate individuals. AI can facilitate communication and organize collective activities, but the design must prioritize inclusivity, dialogue, and collaboration. Flourishing arises when humans intentionally create spaces that encourage interaction, empathy, and shared purpose.

Rituals and traditions play a crucial role in maintaining social cohesion. Celebrations, collective learning, and shared ceremonies anchor community life and transmit values across generations. AI can support the preservation of these practices, documenting histories and simulating experiences, but the lived participation and emotional resonance remain human. Flourishing occurs when communities engage in shared rituals, fostering belonging, identity, and intergenerational continuity.

Equity and inclusion are fundamental to thriving communities. Marginalized groups, whether defined by race, socioeconomic status, disability, or other factors, must have voice and representation. AI can be a tool to amplify inclusion, identifying disparities and creating opportunities, but humans must ensure that decision-making remains just and participatory. Flourishing arises when communities intentionally design systems and structures that value every member's contribution and dignity.

Storytelling and shared narratives strengthen moral imagination. Collective stories convey lessons, ethics, and aspirations, guiding behavior and shaping societal norms. AI can assist in archiving, analyzing, and sharing stories at scale, yet authentic engagement requires human interpretation, discussion, and reflection. Flourishing occurs when communities co-create and participate in narratives that foster ethical reasoning, empathy, and social cohesion.

Education plays a pivotal role in fostering community and shared meaning. Schools and learning spaces should emphasize collaboration, ethical reflection, and civic responsibility alongside academic skills. AI can facilitate personalized learning, collaborative projects, and global connections, but humans must teach the values, empathy, and relational skills necessary for cohesive communities. Flourishing arises when education balances knowledge acquisition with the cultivation of social, moral, and cultural awareness.

Intergenerational engagement strengthens bonds and preserves collective wisdom. Elders provide perspective, experience, and cultural memory, while younger generations contribute energy, innovation, and fresh insight. AI can help bridge generational divides through communication tools and shared platforms, but meaningful understanding requires human interaction. Flourishing occurs when communities foster mutual respect, learning, and collaboration across ages, ensuring continuity and adaptation.

Civic engagement is essential for shared meaning. Participation in decision-making, local governance, and social initiatives empowers individuals while reinforcing collective responsibility. AI can enhance civic participation by providing information, simulating outcomes, and facilitating discussion, but ethical judgment and active involvement remain human responsibilities. Flourishing arises when humans engage purposefully in civic life, shaping communities through collaboration, accountability, and shared vision.

Digital spaces must be intentionally designed to foster connection, not isolation. Social media, virtual worlds, and AI-mediated platforms can either fragment or unite communities depending on their structure and moderation. Thoughtful design prioritizes dialogue, empathy, and inclusion, countering polarization and superficial engagement. Flourishing occurs when humans curate digital environments that reinforce collaboration, understanding, and shared purpose.

Cultural expression reinforces community and shared meaning. Music, art, literature, and performance provide avenues for collective identity, empathy, and communication. AI can assist in creating, preserving, and disseminating cultural works, yet the emotional resonance and interpretive engagement remain human. Flourishing occurs when humans actively participate in cultural life, using creativity to strengthen bonds and communicate shared values.

Trust is the foundation of strong communities. When individuals believe in the reliability, integrity, and intentions of others, cooperation and collaboration thrive. AI can facilitate transparency and accountability, but the cultivation of trust ultimately depends on human behavior, ethical leadership, and consistent action. Flourishing arises when humans foster trust through honesty, empathy, and shared responsibility, creating resilient social networks.

Conflict resolution is an essential skill for sustaining community. Differences of opinion, values, and interests are inevitable, yet communities flourish when they navigate conflict constructively. AI can mediate dialogue, provide data-driven insights, and model negotiation strategies, but empathy, compromise, and moral judgment are human imperatives. Flourishing occurs when humans resolve disputes thoughtfully, preserving relationships and reinforcing collective cohesion.

Shared purpose drives motivation and engagement. Communities that articulate common goals, values, and aspirations provide direction and meaning for individual and collective action. AI can help coordinate efforts, track progress, and suggest optimizations, but alignment with shared human values requires deliberate human engagement. Flourishing arises when people work together toward collective goals, integrating individual talents into a larger vision.

Concluding the discussion on reclaiming community and shared meaning, it is evident that human flourishing is inseparable from the strength of collective life. AI can enhance communication, coordination, and knowledge sharing, but it cannot replace empathy, moral judgment, and relational engagement. Flourishing arises when humans actively nurture relationships, shared purpose, and collective responsibility, ensuring that technology serves community rather than isolates individuals.

Communities provide a framework for identity, ethical reasoning, and resilience. They help individuals navigate complexity, uncertainty, and social change, offering guidance, support, and shared knowledge. AI can facilitate collaboration and insight, but the bonds that sustain communities are built through human connection, care, and commitment. Flourishing occurs when humans invest in nurturing these bonds and strengthening social cohesion.

Cultural and civic engagement is central to meaningful community life. Participation in creative, educational, and civic endeavors fosters shared understanding and reinforces ethical and moral frameworks. AI can amplify these activities, providing tools for organization, collaboration, and dissemination, but genuine engagement requires human initiative and reflection. Flourishing arises when communities collectively pursue shared goals and values with integrity, empathy, and creativity.

Intergenerational and cross-cultural collaboration strengthens the continuity and adaptability of communities. Wisdom, experience, and knowledge from one generation enrich younger members, while fresh perspectives drive innovation and growth. AI can facilitate connections, but human dialogue, mentorship, and exchange of values are essential. Flourishing occurs when communities maintain continuity and adaptability through inclusive, intergenerational, and culturally aware participation.

Finally, reclaiming community is an ethical imperative. Human life is enriched not merely by efficiency, knowledge, or technological advancement, but by connection, reciprocity, and shared meaning. AI can be a powerful tool to support these goals, but human judgment, moral reflection, and active engagement are essential. Flourishing arises when humans place collective well-being at the center of social life, using technology to strengthen—not replace—the ethical, emotional, and relational fabric of communities.

Author Reflection:

Writing this chapter reinforced that human flourishing is fundamentally social. AI can facilitate coordination, communication, and resource sharing, but it cannot create trust, empathy, or shared purpose. True thriving arises when humans actively nurture communities, uphold ethical values, and engage collectively in creating meaning, ensuring that technology enhances rather than diminishes the bonds that sustain life.

Inspirational Quote:

"*No machine can replace the heart of a community; human connection, shared purpose, and collective care are the foundation of flourishing*." — Anonymous

Chapter . 39

Hope Without Naivety

The future demands realism grounded in possibility.

Hope is essential in navigating an AI-driven world, but it must be tempered by awareness and realism. Blind optimism can overlook ethical dilemmas, inequality, and unintended consequences, while despair can paralyze action and innovation. Flourishing arises when humans cultivate hope that is informed, resilient, and guided by understanding of both challenges and opportunities.

Hope requires engagement with reality. Humans must confront the societal, economic, and ethical implications of AI, recognizing the potential for disruption alongside its possibilities for progress. Optimism grounded in knowledge allows for informed decision-making, proactive problem-solving, and meaningful innovation. Flourishing occurs when hope motivates responsible action, rather than naive expectation, creating pathways for positive impact.

Vision and imagination sustain hope. Humans envision a future that aligns with values, ethics, and human well-being, imagining possibilities that AI alone cannot generate. Creative foresight inspires initiatives, policies, and innovations that shape AI in ways that enhance society. Flourishing arises when humans combine imagination with critical thinking, translating vision into tangible actions and structures.

Hope without naivety also requires resilience. Challenges, setbacks, and unintended consequences are inevitable in the integration of AI into society. Human communities must develop adaptive capacity, moral courage, and emotional endurance to respond effectively. Flourishing occurs when hope equips individuals and societies to persist, recover, and grow amidst uncertainty and change.

Hope without naivety relies on informed participation. Citizens, policymakers, and developers must engage deeply with AI technologies, understanding their capabilities, limitations, and ethical implications. Active involvement prevents misuse, amplifies benefits, and cultivates societal trust. Flourishing occurs when humans transform awareness into action, shaping the future intentionally rather than passively accepting technological outcomes.

Ethical foresight is central to sustaining hope. Anticipating potential harms, such as bias, inequity, and environmental consequences, allows humans to design safeguards and interventions proactively. AI offers tools for predictive modeling and scenario planning, but human judgment, values, and moral imagination guide ethical implementation. Flourishing arises when hope motivates ethical foresight, balancing optimism with responsibility.

Collaboration strengthens realistic hope. Global cooperation, interdisciplinary teams, and cross-sector partnerships expand perspectives, identify risks, and generate solutions that no single actor could achieve alone. AI can facilitate collaboration by providing insights, communication tools, and coordination, but human empathy, negotiation, and shared purpose remain indispensable. Flourishing occurs when collective effort channels hope into meaningful, equitable, and sustainable outcomes.

Hope is strengthened by adaptability. In a rapidly evolving technological landscape, flexibility and learning are essential. Humans must be willing to revise assumptions, adopt new strategies, and pivot in response to emerging challenges and opportunities. Flourishing arises when hope is dynamic, allowing humans to maintain vision while adjusting to complex, unpredictable realities.

Hope without naivety depends on critical discernment. In a world flooded with data, predictions, and algorithmic suggestions, humans must evaluate information carefully, identifying both opportunities and risks. AI can support analysis and provide insights, but judgment, context, and ethical evaluation remain human responsibilities. Flourishing arises when hope is informed, guiding decisions that are thoughtful, responsible, and purpose-driven.

Long-term thinking strengthens hope. Immediate gains and short-term metrics often dominate AI development, but sustainable flourishing requires consideration of future generations, societal impact, and environmental consequences. Humans must envision outcomes beyond efficiency and profit, prioritizing the well-being of people and the planet. Flourishing occurs when hope integrates long-term ethical considerations into present action, balancing progress with stewardship.

Hope thrives on creativity and innovation. Humans imagine new ways to solve pressing problems, leveraging AI as a tool to expand possibilities. Novel approaches to health, education, social justice, and environmental sustainability emerge when hope inspires experimentation and courage. Flourishing arises when humans harness imaginative capacity alongside technological innovation, creating solutions that enhance quality of life and collective well-being.

Equity and inclusion anchor hope in reality. Optimism without attention to systemic disparities risks reinforcing existing inequities. AI must be guided to benefit diverse communities, reduce barriers, and amplify marginalized voices. Flourishing occurs when hope is paired with intentional action to create a fairer, more inclusive society where opportunities are shared broadly, not concentrated narrowly.

Hope without naivety depends on ethical leadership. Leaders must navigate the complexities of AI integration, balancing innovation, societal well-being, and moral responsibility. Visionary guidance rooted in humility, empathy, and accountability inspires collective action and safeguards against misuse or unintended harm. Flourishing arises when leaders embody hope grounded in responsibility, modeling ethical engagement for their communities.

Transparency strengthens hope by building trust. Openness about AI's capabilities, limitations, and decision-making processes enables society to participate meaningfully in shaping outcomes. Without transparency, optimism can become misplaced or manipulative, undermining collective confidence. Flourishing occurs when humans combine hope with clarity, ensuring that aspirations are supported by honest information and shared understanding.

Hope is reinforced by community and collaboration. Shared optimism motivates collective problem-solving, strengthens resilience, and fosters solidarity. Humans working together, rather than in isolation, can address challenges that AI alone cannot solve, creating ethical, inclusive, and sustainable solutions. Flourishing arises when collective hope drives collaborative action, reinforcing the social bonds necessary for thriving communities.

Reflection and mindfulness sustain realistic hope. In the rush of technological change, pausing to consider implications, values, and priorities prevents reactive decisions and ethical oversights. Thoughtful reflection allows humans to evaluate progress, adjust strategies, and integrate lessons from experience. Flourishing occurs when humans cultivate reflection as a companion to hope, ensuring that optimism is informed and purposeful.

Concluding the discussion on hope without naivety, it is clear that optimism must be balanced with knowledge, ethics, and realism. AI presents both extraordinary opportunities and profound challenges, and human flourishing depends on navigating this tension wisely. Flourishing arises when hope inspires action grounded in ethical responsibility, critical reflection, and collective engagement, ensuring that optimism leads to meaningful outcomes.

Hope is both personal and collective. Individual aspiration motivates creativity, learning, and resilience, while shared hope strengthens communities, fosters collaboration, and amplifies impact. AI can facilitate communication, insight, and coordination, but human judgment, empathy, and moral commitment are essential to translate possibility into positive change. Flourishing occurs when humans cultivate hope as a guiding force for ethical and socially responsible innovation.

Ethical foresight anchors hope in reality. Anticipating risks, evaluating consequences, and implementing safeguards prevent optimism from becoming blind faith. AI can support predictive analysis, but human discernment ensures that decisions are just, equitable, and sustainable. Flourishing arises when hope is informed, deliberate, and grounded in ethical awareness, creating resilience in the face of uncertainty.

Adaptability reinforces hope. Technological, societal, and environmental landscapes evolve rapidly, demanding flexibility, learning, and continuous engagement. Humans must embrace change while maintaining core values, ethical principles, and shared vision. Flourishing occurs when hope is dynamic, equipping people to respond constructively to emerging challenges while pursuing long-term collective well-being.

Finally, hope without naivety inspires purposeful action. Belief in positive outcomes motivates humans to innovate, collaborate, and act responsibly, shaping AI to serve human and societal good. Flourishing arises when hope is not passive but catalytic, empowering humans to create a future that balances technological advancement with ethical, social, and emotional integrity.

Author Reflection:

Writing this chapter reminded me that hope is most powerful when it is informed, ethical, and active. In an AI-driven world, optimism alone is insufficient; it must be coupled with discernment, responsibility, and collaboration. True human flourishing emerges when hope motivates thoughtful action that enhances collective well-being and aligns technology with human values.

Inspirational Quote:

"*Hope is not wishful thinking; it is informed action guided by ethics, courage, and collective responsibility.*" — Anonymous

Chapter . 40

Becoming More Human Than Ever

AI's greatest gift may be forcing humanity to rediscover itself.

Artificial intelligence challenges the very essence of what it means to be human. As machines assume tasks once thought uniquely human—creativity, analysis, and decision-making—humans are compelled to reconsider their identity, purpose, and value. Flourishing arises when AI prompts reflection on human distinctiveness, emphasizing empathy, ethical judgment, imagination, and emotional depth.

AI accelerates the shift from human labor to human meaning. With many tasks automated, humans have the opportunity to focus on what machines cannot replicate: relationships, moral reasoning, self-awareness, and curiosity. Flourishing occurs when individuals embrace this transition, cultivating capacities that are uniquely human while leveraging AI as a tool to extend their potential.

Reflection and self-awareness are central to becoming more human than ever. Rapid technological change demands that people pause, assess their values, and clarify their purpose. AI can support self-discovery through personalized insights and feedback, but genuine understanding arises from introspection, moral discernment, and lived experience. Flourishing occurs when humans actively cultivate self-knowledge, ethical clarity, and authentic engagement with the world.

Creativity and imagination flourish in partnership with AI. While machines can generate ideas and suggest solutions, humans provide context, meaning, and emotional resonance. AI expands the canvas for human expression but does not replace the capacity to wonder, dream, and innovate. Flourishing arises when humans use technology to amplify creative potential, preserving the uniquely human ability to imagine, interpret, and transform.

Human identity evolves in response to AI's capabilities. As machines perform analytical, creative, and decision-making tasks, humans are freed to explore higher-order dimensions of thought, emotion, and morality. Flourishing arises when humans embrace this evolution, cultivating self-expression, ethical reasoning, and curiosity as central aspects of identity.

The cultivation of wisdom becomes increasingly essential. In a world of instant answers and algorithmically generated solutions, discernment, judgment, and reflection distinguish humans from machines. AI can process data efficiently, but wisdom emerges from experience, moral insight, and contextual understanding. Flourishing occurs when humans prioritize wisdom, integrating knowledge, empathy, and foresight in guiding actions and relationships.

Purpose becomes a core human pursuit. Work, once tied to survival and status, may no longer define identity or meaning. AI enables humans to pursue endeavors aligned with personal values, social contribution, and creative exploration. Flourishing arises when individuals identify and engage in purposeful activities that nurture growth, connection, and fulfillment beyond economic necessity.

Ethical responsibility is magnified in an AI-augmented world. Human actions increasingly shape the social, economic, and environmental impact of intelligent systems. Flourishing occurs when humans embrace moral accountability, ensuring that AI development and use reflect compassion, justice, and shared well-being.

Learning and curiosity remain distinctly human frontiers. AI can supply knowledge, analyze patterns, and generate insights, but the drive to ask questions, explore unknowns, and challenge assumptions is uniquely human. Flourishing arises when individuals cultivate lifelong curiosity, using AI as a tool to deepen understanding rather than as a substitute for inquiry.

Emotional intelligence becomes a defining human asset. The capacity to recognize, manage, and empathize with feelings—both one's own and others'—cannot be replicated by machines. Flourishing occurs when humans invest in emotional awareness, resilience, and compassionate action, ensuring that relationships and social systems thrive in tandem with technological progress.

Collaboration with AI encourages humans to think critically about their choices and values. Machines provide analysis, recommendations, and automation, but humans must interpret, contextualize, and make final decisions. Flourishing arises when humans embrace the responsibility of guiding AI ethically and thoughtfully, using technology to enhance rather than replace judgment and conscience.

Creativity gains new dimensions through human-machine partnerships. AI can generate ideas, simulate scenarios, and assist with experimentation, yet the integration of meaning, aesthetic sensibility, and cultural resonance remains a human prerogative. Flourishing occurs when humans harness AI to amplify creative potential while preserving the uniquely human touch of imagination, narrative, and interpretation.

Spiritual and existential exploration becomes increasingly important as AI reshapes human experience. Machines can simulate understanding, provide information, and even generate art, but they cannot grapple with questions of meaning, purpose, and transcendence. Flourishing arises when humans engage deeply with spiritual and philosophical inquiry, using technology to support, not replace, their search for understanding.

Freedom and autonomy are essential human qualities in an AI-enhanced world. Automation may relieve humans of routine labor, yet the responsibility to make ethical choices, pursue meaningful goals, and shape society remains. Flourishing occurs when individuals exercise agency, balancing the convenience AI provides with conscious, value-driven decision-making.

Resilience and adaptability are strengthened through conscious engagement with technological change. Rapid advancements challenge individuals and societies to learn continuously, reassess priorities, and navigate uncertainty. Flourishing arises when humans embrace change as an opportunity for growth, integrating AI into life thoughtfully and ethically rather than reacting passively to disruption.

Interconnectedness highlights humanity's evolving role. AI connects people, societies, and knowledge in unprecedented ways, revealing the interdependence of human actions and technological systems. Flourishing occurs when humans recognize this interconnection, collaborating across boundaries, nurturing communities, and taking responsibility for the collective impact of their decisions.

Concluding the discussion on becoming more human than ever, it is clear that AI's rise is not a threat to humanity but an invitation to rediscover and amplify human distinctiveness. Machines can perform tasks, generate insights, and simulate intelligence, but the qualities that define humans—empathy, moral judgment, creativity, reflection, and purpose—cannot be replicated. Flourishing arises when humans embrace these capacities intentionally, using AI to extend their potential rather than replace it.

Ethical responsibility becomes central in this journey. As humans shape and deploy intelligent systems, their choices carry profound consequences for society, the environment, and future generations. Flourishing occurs when humans act with awareness, integrating ethical principles, foresight, and empathy into every decision that affects the collective well-being.

Reflection and self-awareness anchor human identity. In the face of rapid automation and AI proliferation, taking time to examine values, aspirations, and purpose allows humans to navigate change with integrity. Flourishing arises when humans cultivate intentionality and mindfulness, ensuring that technological progress enhances rather than erodes the human spirit.

Collaboration with AI magnifies human capabilities while reinforcing uniquely human roles. Machines can process information, suggest solutions, and optimize processes, but interpretation, meaning-making, and ethical judgment remain human domains. Flourishing occurs when humans leverage AI thoughtfully, fostering creativity, insight, and moral clarity.

Finally, the ultimate promise of AI is to illuminate what it means to be fully human. By delegating routine, computational, and repetitive tasks, humans are freed to explore relationships, ethics, creativity, and purpose at a depth previously unattainable. Flourishing arises when humans embrace this opportunity, cultivating the qualities that make life meaningful, ethical, and profoundly human in an increasingly intelligent world.

Inspirational Quote:

"AI may outthink us, but only humans can love, reflect, and create meaning; our greatest power is to become more human than ever."
— Anonymous

Epilogue:

The Choice to Flourish

As this book concludes, one truth resonates: AI is not destiny—it is a tool, a challenge, and an invitation. The machines we build reflect our values, priorities, and imagination. They amplify both our strengths and our shortcomings. The question we face is not whether AI will change the world, but whether we will rise to the occasion and ensure that it changes us for the better.

Human flourishing in the age of AI is both personal and collective. It requires reflection, empathy, ethical judgment, creativity, and curiosity. It demands attention to relationships, community, and shared meaning. It challenges us to act with responsibility and courage, to balance optimism with realism, and to design systems that elevate human life rather than diminish it.

The future is not predetermined. Each generation has faced transformative technologies and has made choices—sometimes wisely, sometimes poorly—about how to integrate them into society. Now, as intelligent machines proliferate, we have the opportunity to act deliberately, ethically, and creatively. Flourishing is a choice: it is the conscious effort to cultivate what is uniquely human while leveraging technology to extend our potential.

The next steps belong to all of us. Parents, educators, policymakers, workers, creators, and citizens are called to participate in shaping AI in ways that honor human dignity, creativity, and collaboration. By embracing hope without naivety, slowing down to reflect, reclaiming community, and cultivating ethical awareness, we can ensure that AI serves humanity, not the other way around.

Above all, the promise of this era is not simply technological progress—it is a renewed opportunity to become more human than ever. Machines may think, calculate, and optimize, but humans can love, imagine, create, reflect, and act with moral purpose. The question remains: will we rise to meet that opportunity? The choice is ours, and in that choice lies the future of human flourishing.

About the Author:

Laura Lee is a writer, creative entrepreneur, and digital visionary whose work bridges imagination, heart, and innovation. Growing up in a quiet town where stories were lifelines, she discovered early the transformative power of expression and the deep connections that words and ideas can create.

As the founder of a thriving digital design agency, Laura has guided countless brands in shaping identities that are clear, purposeful, and infused with meaning. Her writing—intimate, reflective, and universally resonant—echoes the same values that guide her work: authenticity, conscious creation, and emotional depth.

For Laura, design is never just about aesthetics—it is about intention. Love, in all its forms, is never just about romance—it is about truth. Through her books, digital platforms, and creative consulting, she inspires others to find their voice, honor their journey, and cultivate lives rooted in integrity, beauty, and purpose.

This book is a natural extension of that mission: a heartfelt guide to self-discovery, a reminder to pause, and an invitation to reclaim meaning in a world that often asks us to forget who we are.

www.ingramcontent.com/pod-product-compliance
Lightning Source LLC
LaVergne TN
LVHW052300200926
848481LV00002B/425